ADAPTIVE TECHNIQUES FOR MIXED SIGNAL SYSTEM ON CHIP

The International Series in Engineering and Computer Science

VOLUME 872

Editors

Adaptive Techniques for Mixed Signal System on Chip

by

Ayman Fayed
Texas Instruments Inc., Dallas, Texas, U.S.A.

and

Mohammed Ismail
Analog VLSI Lab, Ohio State University Columbus, Ohio, USA

A C.I.P. Catalogue record for this book is available from the Library of Congress.

ISBN-13 978-1-4419-4071-1
ISBN-13 978-0-387-32155-4 (e-book)

Published by Springer,
P.O. Box 17, 3300 AA Dordrecht, The Netherlands.

www.springer.com

Printed on acid-free paper

Dedication

To Layla, Lobna, my parents,
and my country Egypt.

And To Ismail, Jr., Omar,
Tuula, and to Sameha and
Ismail, Sr.

Contents

Preface

Analog and mixed signal integrated systems of today and tomorrow will be very complex as they meet the challenge and increased demand for higher levels of integration in a System on Chip (SoC). Current and future trends call for pushing system integration to the highest levels in order to achieve low cost and low power for large volume products in the consumer and telecom markets, such as feature-rich handheld battery-operated devices. While CMOS technology scaling to nanometer levels, coupled with innovations in platform based systems and Network-on-Chip (NoC) have resulted in great strides with the digital part of a system, the analog, radio, or mixed signal part of the total solution remains a major bottleneck. In today's analog design environment, a fully integrated CMOS integrated circuit may require several silicon spins before it meets all product specifications and often with relatively low yields. This, results in significant increase in NRE (Non Re-occurring Engineering) cost, especially that mask set costs increase exponentially as feature size scales down. Furthermore, this could lead to missing important market windows, particularly with the decreasing life cycles of semiconductor products.

Integrated analog, RF, and mixed signal systems lack arbitrary composability, i.e. they can not be composed from their sub-systems as easily as their digital counterparts. The performance is highly susceptible to random variations in process and operating conditions. Such variations do not scale with the process. Worst-case corner simulations often lead to over-design and increased power consumption. Device and process models, package models, and design kits are based on certain assumptions that severely limit design space exploration. All these factors prohibit first-time-right silicon.

Current design practices simply do not lend themselves to fully integrated mixed signal SoC solutions in emerging nanometer processes.

This book is devoted to the subject of adaptive techniques for smart analog and mixed signal design whereby fully functional first-pass silicon is achievable. To our knowledge, this is the first book devoted to this subject. The techniques described should lead to quantum improvement in design productivity of complex analog, RF and mixed signal systems while significantly cutting the spiraling NRE costs of product development in emerging nanometer technologies. The underlying principles and design techniques presented are generic and would certainly apply to CMOS analog and mixed signal platforms in high volume , low cost wireless ,wire line and consumer electronic SoC or chip set solutions.

The approach we adopt in the book is based on introducing self-awareness and adaptability both locally to circuit blocks or sub-systems and globally at the system level. We re-think the analog design problem in terms of building reliable, predictable systems out of stochastically behaving blocks or sub-systems. Our vision here borrows from organic computing where a well-defined system behavior can emerge from seemingly random interaction of a large number of components. This would target random variation in the system or in the transmission media. Moreover, a basic level of self-awareness is introduced at the block or sub-system level. In a bottom-up approach, sources of errors in each block are recognized and corrected. Self-tuning adaptive techniques that could be analog, digital or mixed signal are then applied. Examples and case studies of self-awareness, tunable elements as well as on chip digital and analog adaptive techniques are presented.

Chapter 1 discusses the essence and process of adaptation in the context of analog and mixed signal design and provides a complete outline of the book. Chapter 2 introduces different adaptive architectures used in control of any system parameter including absolute values of on chip resistors or capacitors. It also gives an overview of the different elements used including tunable elements as well as voltage, current and time reference circuits. Chapter 3 focuses on design of tunable elements such as transconductors and offset comparators, both being voltage-controlled elements. Chapter 4 gives an overview of the sources of errors in components due to random process and temperature variations while chapter 5 describes a digital adaptive technique for the accurate implementation of on chip resistors. A novel adaptive algorithm is used in this work. While these chapters address adaptive techniques at the circuit and block levels, Chapters 6 and 7 discuss errors in system and transmission media and introduce equalization architectures that could be analog, digital or mixed signal. The material focuses on minimizing the impact of ISI (Inter symbol Interference) on the quality of received data and presents adaptive equalization architectures to implement a 125Mbps transceiver operating over a variable length of up to 100 meters of an Unshielded-Twisted-Pair (UTP) Category-5 (CAT-5) Ethernet cable.

The book is intended for use by graduate students in electrical and computer engineering as well as analog, RF, and mixed signal design engineers in the semiconductor and telecom industries. It will also be useful for design managers, project leaders and individuals in marketing and business development.

This book has its roots in the Ph.D. thesis of the first author. We would like to thank all those who assisted us at different phases of this work including our colleagues at the Analog VLSI Lab, The Ohio State University and at the Connectivity Solutions Department at Texas Instruments. Technical discussions with M. Al-Shyoukh, M. Kamal, R. Griffith, and J. Wilson are also acknowledged. We would like to thank our families for their understanding and support during the development of this work.

Ayman Fayed, Dallas, Texas
Mohammed Ismail, Columbus, Ohio

Chapter 1

INTRODUCTION

Implementing modern electronic systems with mixed analog and digital content has become a very challenging task due to the strong drive for reducing cost, increasing system speed, and widening the range of conditions a specific design can support and tolerate. Reducing cost essentially implies minimizing the area of the printed circuit board that carries the system, minimizing the number and price of individual ICs that constitute the system, and minimizing the cost of the package that carries each IC. In other words, reducing cost implies achieving the highest level of integration using the cheapest possible packaging technology without compromising speed or performance. From purely digital systems perspective, this led to the development of deep sub-micron digital CMOS technologies that can accommodate very large sophisticated digital designs with very small die area, low power consumption, and high speed performance. Yet from a mixed signal systems perspective, where heavy analog content needs to be implemented along with the digital content, this introduces many limitations on what analog circuit designers have at their disposal for implementing the required functionalities using the same digital CMOS technology. For instance, decreasing the feature size of the MOS transistor to reduce die area, as well as decreasing its threshold voltage and its gate capacitance in order to increase its switching speed capability implies increasing its gate leakage current as well as reducing the maximum voltage level the transistor can handle. On the other hand, as the feature size of the transistor decreases, the control over the absolute values of its parameters becomes even less. This makes it very difficult to predict the behavior of the transistor and consequently the analog circuits that use it. Even, in the best case, if the behavior of the transistor is well predicted and modeled, the possibly large and random variations in the transistor's characteristics, the lower supply voltage the transistor can handle, and the higher gate leakage current makes the design of even simple analog circuits rather a very complicated task. This is in addition to the fact that the transistor's lower voltage capability makes it even more difficult to interface with the outside world that could potentially have higher voltage levels than the transistor can handle. On top of that, the poor isolation in these technologies and the lack of twin tubs introduces issues like substrate noise coupling and body effects. Those are just the challenges introduced by the process technology, but in addition to that, minimizing the cost of packaging introduces its own set of challenges as well, where low cost packages usually exhibit higher parasitics that make the design of analog circuits that interface with the outside world even more challenging. Given all those limitations though, analog circuit designers have to still meet the very tight specifications required for adequate performance of modern sophisticated systems.

All the previous factors combined makes the design process very cumbersome and forces the development of more sophisticated circuits to perform functions that used to be thought of as very simple and could be implemented with guaranteed first-pass silicon success. The need for more sophisticated circuits and the possibly larger variations in the transistor's characteristics automatically imply a higher failure rate in meeting the required specifications, or at best, a significantly lower yield. Consequently, it usually takes multiple silicon spins to get the performance of the circuit to where it needs to be, which makes "first-pass silicon success" very difficult to achieve. Obviously, this multiple spins process means a much higher cost, which adds up to the original high cost of modern fabrication processes (almost a million dollars today for an all-level mask set in a 90nm process).

1.1 Categories of Variations in Analog and Mixed Signal ICs

The previous discussion essentially implies that circuit designers have to achieve the required performance using devices that have wide random variations in their characteristics, and under conditions that could potentially vary significantly during operation. This means that the design needs to meet the targeted specification with a significant margin. Variations in a typical analog and mixed signal system could be categorized into three main categories; circuit level, system level, and network level.

1.1.1 Circuit Level Variations

Circuit level variations are encountered during the design of each individual circuit block in a system. Those variations include random process variations, supply voltage variations, and temperature variations.

Process variations are caused by the limited control over the conditions and characteristics of the fabrication process. This limited control causes variations in the length, width, and threshold voltage of transistors as well as the absolute values of on-chip resistors and capacitors. For example, an on-chip resistor could vary by as much as ±25% from its targeted nominal value just because of variations in the fabrication process parameters. In addition to random process variations, the system has to meet the required set of specifications for a range of supply voltages as well as temperatures. Typically, power supplies are generated using voltage regulators (switching, or linear regulators) that regulate an input power source (a battery or an AC source) to the voltage required by the system. The accuracy of this regulation process is a function of the accuracy of the reference used, variations in the input power source voltage, and variations in the load current. The typical supply voltage range is usually ±10% of the targeted nominal value, even though more accurate supplies are achievable for a higher cost. Since the ambient and junction temperatures in any circuit could widely vary depending on the operation conditions of the circuit, the circuit has to be able to operate adequately across a range of temperatures. The typical temperature range for consumer electronics applications is from 0 to 80 °C, while for military and automotive applications, it could be as wide as from -40 to 125 °C. Taking all those variations into account, it becomes a challenging task to design circuits that can accurately meet the tight specifications set on a system with all those variations present. For instance, in a digital CMOS process, the design of continuous-time filters with on-chip resistors and capacitors becomes impractical for many applications since the values of those resistors and capacitors could vary by as much as ±25% from their intended nominal value. This makes it very difficult to control the positions of poles and zeros in a given filter. Another example is the implementation of integrated transmission line termination resistors for digital high speed wire line communication systems. In this case, the variations in those resistors from their nominal value will cause mismatches with the transmission line characteristic impedance, which in turn will cause reflections, and hence loss of signal power and integrity[1-4]. Generally speaking, the design of circuits that require accurate resistors, capacitors, voltages, or transistor parameters becomes a difficult task. This includes filters, oscillators, delay elements, transmission lines terminations, and so on. Even in circuits that wouldn't need resistors and capacitors, the variations in the mobility and threshold voltage of transistors as well as the length and width could still be a challenging problem. For example, signal drivers in wire line transceivers usually have restrictions on the rise and fall times as well as duty cycle distortion. The system specifications usually impose an upper and lower limits on those parameters (usually given in an eye diagram format). If the range of rise and fall times allowed by the specifications is tight, it becomes very difficult to meet those specifications given the variations in the supply voltage, junction temperature, and transistor parameters.

Different techniques have been developed in the literature to avoid the dependency on the absolute values of device parameters in general. One technique is based on developing circuits for implementing the required system (filter, oscillator ...etc.) such that the targeted parameter (cut-off frequency, oscillation frequency ...etc.) of the system would depend on a ratio between parameters of similar on-chip elements (resistors, capacitors, or transistors) rather than the absolute value of their individual parameters (Switched-Capacitor techniques is an example[5]). As will be discussed later in detail, the ratio between parameters of similar on-chip elements has a much better accuracy than their individual absolute values. Unfortunately though, it is not always possible to implement the required system such that the parameter of interest depends solely on a ratio between similar elements. It is inevitable in these cases to depend on the absolute value of some on-chip element parameter.

1.1.2 System level variations

System level variations are related to the specifications and type of the system, i.e. variations that cause the specifications of the system to be changeable. For example, there is more than one mobile phone standard (GSM, TDMA …etc.). The need for a single transceiver that can automatically handle different standards has been increasing recently. This means that there will be two completely different sets of specifications that the transceiver has to handle. The impact of having two different sets of specifications on the circuit level is very significant. There are two options: designers can either use two separate circuit blocks with each block handling one set of specifications and switch between them, or one circuit block that can handle both sets of specifications. Obviously, the first option may not be desirable since it will consume almost double the area even though it might be a logical starting point. The second option adds another design challenge in addition to the one posed by the variations in the circuit level. For example, designers have to find a way to design a filter that can automatically switch between two different specifications on the bandwidth and center frequency. This applies also to oscillators, low noise amplifiers, analog to digital converters, and so on. Therefore, not only do designers have to compensate for process, power supply, and temperature variations to comply with one set of specifications, they also have to comply with different sets of system as well as circuit specifications.

1.1.3 Network level Variations

Network level variations are mainly variations in the characteristics of transmission media. There are two different types of networks: wire line networks, and wireless networks. Usually the two biggest problems introduced to digital communication systems through the transmission media is Intersymbol interference (ISI) and cross channel interference. ISI is generally caused by the limited bandwidth of the transmission media. The limited bandwidth causes each transmitted symbol to be spread in time beyond its allocated time slot, consequently, interfering with other symbols in the same data stream. In wireless networks, and in addition to limited bandwidth, ISI could also happen due to the multi-path effect, in which an indirect delayed signal that usually results from multiple reflections interferes with the direct signal received by the antenna causing the current received symbol to be contaminated with previous symbols in the same channel. In wire line networks, ISI is a function of cable length and characteristics. For example, in Ethernet networks, the cable could be of any length up to hundred meters. For each length, ISI is different due to the different frequency response of each cable length. For both wire line and wireless systems, equalizers are employed to resolve the ISI problem, but designers are challenged by the fact that not only do they need to equalize for a fixed channel, they also need to equalize for a channel that can change characteristics in an unpredictable way.

1.2 The Essence of Adaptation

Considering the variations on the circuit, system, and network levels discussed in the previous section, it is obvious that the common challenge is the problem of designing circuits that can handle those variations, preferably in an automatic way. Whether the variations are in the elements composing the circuits (circuit level variations), or in the specifications the circuits have to meet (system level variations), or in the transmission media characteristics (network level variations), the common solution is adaptation. Adaptation is a very broad concept that can be applied to so many fields in the area of analog and mixed signal circuit design. The first step in any adaptive scheme is to sense the parameter of interest, followed by the second step, which is comparing the sensed parameter to an accurate reference resulting in an error signal that indicates the amount of variation in the parameter of interest. The third step involves further processing of the error signal according to a specific algorithm, which results in a control signal that could then be used to either directly control the parameter of interest in order to minimize or perhaps eliminate the variation represented by the error signal, or to control other different elements in the system to compensate for those variations and keep the performance within the required specifications. Circuits and systems that employ adaptation techniques are called adaptive circuits and systems. Adaptive equalization, automatic impedance control, time constant control, offset cancellation techniques, and multi-mode systems are just few examples of systems that apply adaptation techniques in one way or another.

1.3 The Process of Adaptation

Adaptation techniques are based on sensing the variation in a specific parameter in the system and then quantify this variation. Based on the variation amount, the rest of the system adapts itself to compensate for this variation. Taking that into consideration, using adaptive techniques requires two levels of innovation, i.e. the circuit level and the system level. On the circuit level, innovation includes the design of circuits that enables adaptation, while on the system level, innovation includes the different algorithms or architectures used to perform the adaptation process. As mentioned previously, in order to adapt for a variation in a specific parameter, the parameter has to be sensed. This will require the design of a circuit that can track and sense wide variations in that parameter. Therefore, the sensing circuit has to have a wide input and output range and it also has to be insensitive as much as possible to process and system variations in order to be used efficiently in the sensing process. Sensing circuits might be amplifiers, op-amps, transconductors, filters and so forth. The output of the sensing circuit has to be then compared to a reference signal, which implies that comparison techniques have to be developed. Those comparison techniques have to be able to handle wide input range and might have to have a highly linear performance for adequate comparison. In the analog domain, the comparison could simply be done using a differential amplifier, while in the digital domain, analog to digital converters may have to be used.

Any comparison technique needs some sort of a reference signal, and in order for the comparison result to be meaningful, this reference has to be insensitive to all the variations and the parameters that the system is trying to detect and compensate for. Designing an accurate reference could be a very difficult and challenging task, which leads to another level of innovation as will be discussed later in details. The result of the comparison process is an error signal that gets processed further to generate a control signal that is used to tune the parameter of interest back to its desired value. This tuning process introduces another challenge, i.e. the design of circuits that have variable, or in other words, programmable characteristics that can be controlled with a control signal (a voltage or a current signal). This includes the design of variable gain amplifiers, voltage controlled transconductors, programmable filters, programmable resistors, programmable time constants, and so on. Those circuits have to have a large programmable range (perhaps linear as well) in order to adapt for large variations in the parameter of interest. Taking into account the sensing process, the comparison process, and the programmable behavior of circuits, using adaptive techniques open a wide range of possibilities for circuit level innovations.

The system level innovation includes two elements: the algorithm used for processing the error signal in order to generate the control signal, and the architecture of the adaptation process in general. The algorithms used for processing the error signal are very important since they determine the total accuracy, speed, and stability of the adaptation process. Heavy mathematical analysis has to be performed in order to develop the best and most efficient algorithm for a given application.

There are generally three possible techniques that could be used to implement an adaptive system, i.e. analog techniques (continuous or discrete), digital techniques, or mixed signal techniques. The error signal that results from the comparison process could be an analog signal (continuous or discrete) or a digital word. Based on the nature of the error signal, the further processing needed on it in order to generate the control signal could be done either in the digital or the analog domain. Once the control signal is generated and based on its nature being analog or digital, the circuits in the system could be programmed by either analog or digital means respectively. Therefore, the adaptation process could be done purely in the analog or the digital domain, or it can employ both techniques simultaneously. Digital techniques are very powerful since they enable very complex algorithms to be performed using digital signal processors (DSP), but they usually require analog to digital converters (ADC) and digital to analog converters (DAC). Particularly, if the variations in the parameter of interest are much larger than its desired accuracy, the design of the ADC becomes very difficult since high resolution is needed. Analog techniques on the other hand don't have the resolution limitation, but mathematical processing required on the error signal becomes relatively complicated to implement in the analog domain. Mixed signal architectures are usually used to achieve the advantages of both analog and digital techniques. No matter what technique or architecture is used for the adaptation process, stability, convergence time, and speed have to be considered when developing an adaptive architecture or algorithm.

1.4 Book Outline

As mentioned before, three categories of variations can be encountered in analog and mixed signal circuits and systems design, i.e. circuit level, system level, and network level. This book focuses on adaptation techniques for variations on the circuit and network levels only, i.e. variations due to process, supply voltage, temperature (aka PVT), and transmission media. Since using adaptive techniques is very important and sometimes essential for meeting the tight specifications of today's systems, chapter 2 gives a basic introduction to different adaptive architectures that could be used to control the accuracy of the absolute value of any parameter in the system including the absolute values of on-chip resistors and capacitors. An overview of the different elements used in automatic adaptive architectures including references and electronically tunable elements is also presented. This includes voltage, current, and time references, basic circuit techniques for implementing tunable resistors, capacitors, and transconductors. Post fabrication tuning options are also briefly described.

Since tunable elements are necessary for any adaptive architecture to operate successfully, chapter 3 focuses on the circuit design of a high performance voltage-controlled transconductor and a voltage-controlled offset comparator. These two blocks provide wide input/control ranges that enable more robust adaptive architectures.

As discussed before, the absolute value of different parameters of on-chip elements (resistor, capacitors, and transistors) usually suffer from wide variations due to process and temperature inaccuracies. In order to understand and quantify those variations, chapter 4 gives a case study of available types of on-chip resistors and capacitors, followed by a discussion of sources of errors in their absolute values due to variations in the fabrication process as well as temperature. Design and layout techniques to minimize those errors will also be presented. Since matching properties of on-chip resistors and capacitors is a very important aspect of their implementation, chapter 4 will also give an overview of those matching properties with a discussion of the different layout techniques to improve and achieve accurate matching.

Once different adaptive tuning architectures and the elements used to implement those architectures are presented in chapter 2 and 3, as well as sources of error in the absolute value of on-chip resistors and capacitors in chapter 4, chapter 5 proceeds with presenting an application example of an adaptive architecture used for counteracting variations on the circuit level. Specifically, an adaptive tuning architecture for implementing accurate on-chip resistors will be presented. The chapter focuses on the theoretical and mathematical background behind the architecture along with a discussion of its advantages over other widely used techniques. The significance of this part of the book is that it introduces a simple, yet powerful methodology for implementing accurate on-chip resistors. The technique was successfully used for integrating accurate on-chip transmission line termination resistors for a high-speed wire line digital communication transceiver.

Chapters 4 and 5 present a case study as well as an application for compensating for variations on the circuit level, particularly, errors in on-chip resistors. Chapter 6 on the other hand focuses on network level variations, i.e. variations in the transmission media. The chapter discusses the effects of transmission media on the quality of received digital data, particularly Intersymbol Interference (ISI) in digital wireless and wire line communication systems. Available techniques to minimize the impact of ISI on the quality of the received data, namely equalization theory will be discussed. Different equalization architectures and algorithms will also be presented including fixed and adaptive equalization along with the different techniques that could be used (analog, digital, and mixed signal) to implement those architectures. The advantages and disadvantages of each technique will also be discussed.

Using the basic theory presented in chapter 6, chapter 7 presents an application that uses purely analog adaptive equalization architecture to implement a 125Mbps transceiver that operates over a variable length of up to 100 meters of an Unshielded-Twisted-Pair (UTP) Category-5 (CAT-5) Ethernet cable. A discussion of the requirements that every circuit block used in the system needs to meet will be presented, along with simulations and measurements results.

LIST OF REFERENCES

1. H. Conrad, "2.4 Gbit/s CML I/Os with integrated line termination resistors realized in 0.5/spl mu/m BiCMOS technology," *Proceedings of the Bipolar/BiCMOS Circuits and Technology Meeting,* pp. 120-122, Sept. 1997.
2. T.J. Gabara, "On-chip terminating resistors for high-speed ECL-CMOS interfaces," *Proceedings of the Fifth Annual IEEE International ASIC Conference and Exhibit,* pp. 292-295, Sept. 1992.
3. D.R. White, K. Jones, J.M. Williams, I.E. Ramsey "A simple resistance network for calibrating resistance bridges," *IEEE Transactions on Instrumentation and Measurement,* vol. 46, pp. 1068-1074, Oct. 1997.
4. I. Novak, "Modeling, simulation, and measurement considerations of high-speed digital buses," *Instrumentation and measurement Technology Conference,* pp. 1068-1074, May. 1992.
5. David A. Johns, Ken Martin, "Analog Integrated Circuit Design," John Wiley & Sons, New York, 1997.

Chapter 2
ADAPTIVE ARCHITECTURES

The concept of adaptation is a broad concept that can be applied to many fields in the area of analog and mixed signal system and circuit design. As discussed in chapter 1, absolute values of on-chip elements characteristics are highly inaccurate, leading any performance parameter that depends on those characteristics to be highly inaccurate as well. One way to solve this problem is based on developing circuit techniques for implementing the required system (oscillator, filter ...etc.) such that targeted performance parameters (oscillation frequency, bandwidth ...etc.) of the system would depend on a ratio between similar on-chip elements rather than their individual absolute values. As will be discussed later in details, ratios between similar on-chip elements are much more accurate than their individual absolute values. Example of circuits that employ this strategy is switched-capacitor circuits[1]. Unfortunately though, it is not always possible to implement the required system such that the parameter of interest is only a function of ratios between similar elements. In these cases, depending on the absolute values of those elements becomes inevitable, and an adaptive architecture has to be employed to tune out performance degradation resulting from this dependency.

Adaptive flows involve the following fundamental steps: a) sensing the parameter of interest in a given system or circuit, b) comparing the sensed parameter against a precise reference to generate an error signal, and c) processing the error signal according to specific algorithms to generate a control signal that either directly tunes the parameter of interest, or controls other elements in the system to compensate for loss in performance in the parameter of interest. Circuits and systems that employ those adaptive flows are usually referred to as adaptive circuits and systems, or simply tuned circuits and systems. Examples that employ the above methodology include adaptive equalization in high-speed wire line transceivers, automatic gain and impedance control in data transmit/receive paths, and offset cancellation techniques. Adaptive techniques can be categorized into two main categories. The first category is on-chip automatic tuning (self-tuning), and the second category is post fabrication tuning. In this chapter, an overview of different techniques used by both categories and the pros and cons of each one is presented along with a discussion of circuit elements necessary in any adaptive architecture including references and electronically tunable circuits.

2.1 Automatic Tuning

An automatically tuned system is one that has the ability to tune a set of its parameters to an accurate set of values automatically with no need for any external intervention. Thus, it is also referred to as a self-tuned system. Usually there are two methodologies for tuning a parameter in the system. It could be indirectly tuned by tuning the on-chip elements that the parameter is a function of, or it could be directly tuned by other means to compensate for errors in those on-chip elements. The second methodology implies that the parameter of interest has to be a function of another control parameter that can be changed to compensate for any errors. In other words, the parameter of interest has to be electronically tunable. For example, the location of poles in a filter‘s transfer function could be designed to be a function of an amplifier's gain in addition to its dependency on the absolute value of a resistor or a capacitor. By designing the amplifier such that its gain is controlled by a voltage signal, the location of the filter's poles could in turn be adjusted.

The automatic tuning process has three fundamental steps. The first step is measuring the parameter of interest. The second step is error signal generation, which results from comparing the measured value of the parameter of interest to an accurate reference to quantify how off the value of the parameter is from its desired value. The third step is tuning signal generation, in which the error signal is used in conjunction with a negative feedback loop in order to tune the parameter of interest to match the reference. The automatic tuning process could be performed using three different architectures depending on the application. The first architecture is the direct continuous architecture, in which the tuning loop is continuously running during normal operation of the tuned circuit, provided that the tuning loop does not require using the input of the tunable circuit for any purpose. Figure 2-1a shows a conceptual block diagram of the direct continuous tuning architecture.

Very often though, measuring the parameter of interest might require applying a special signal at the input of the tuned circuit that is different in nature than the normal operation input signal. For example, a common way of measuring quality factors of filters is applying a step voltage at the input and counting the number of overshoots and undershoots at the output. In these cases, the input of the tunable circuit has to be disconnected from the normal operation input signal during the tuning period. This leads to the second architecture, which is the direct discontinuous architecture. In this architecture, the tunable circuit has to be taken offline during the tuning process until a control signal value has been reached. The tunable circuit is then put back online with the stored control signal applied to it. The tuning loop is then disabled during normal operation until another tuning procedure is initiated. Figure 2-1b shows a conceptual block diagram of the direct discontinues architecture. Note that if the application does not tolerate taking the tunable circuit offline during the whole tuning period, a copy of the tunable circuit with a previously held control signal can be used temporarily at the input until tuning is over and a new control signal value has been reached. The original tunable circuit can then be put back online for the rest of operation time. However, switching between the original and temporary copies of the tuned circuit implies that the circuit will still be briefly taken offline leaving the input lines disconnected, and the application has to tolerate that. The major advantage of direct architectures (continuous and discontinuous) is that the tuned circuit used in normal operation is the same one being tuned by the tuning loop. Hence, the highest level of tuning accuracy could be achieved using these two architectures.

In some cases when measuring the parameter of interest require applying a special signal at the input of the tuned circuit, but the application does not tolerate taking the tunable circuit offline or disconnecting the input lines even for a short time, then neither the direct continuous or discontinuous architectures will be suitable. In these cases, a master copy of the tunable circuit is used to perform the tuning, and the resulted control signal is then used to tune a slave copy of the tunable circuit, which is continuously connected to the input lines. Since the tuning is performed on a copy of the tunable circuit, this architecture is referred to as indirect tuning architecture. Indirect tuning architectures could also be continuous, where the tuning loop is continuously running, or they could be discontinuous, where the tuning is performed and the resulting control signal is stored. Figure 2-2 shows a conceptual block diagram of the indirect tuning architecture. Note that the master and slave copies of the tunable circuit could be tied together or an excitation signal could be applied only to the master copy, while the slave copy is connected to the input lines. This excitation signal is the special signal required to be applied to the tuned circuit input in order to measure the parameter of interest. The main disadvantage of indirect architectures though is that tuning accuracy is limited by mismatches between the master and slave copies of the tunable circuit.

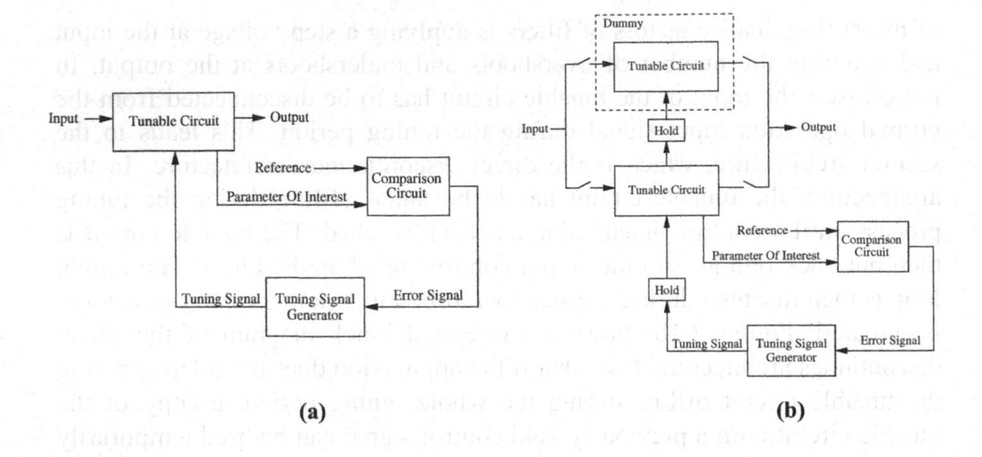

Figure 2-1. Direct tuning architecture: (a) continuous tuning, (b) discontinuous tuning

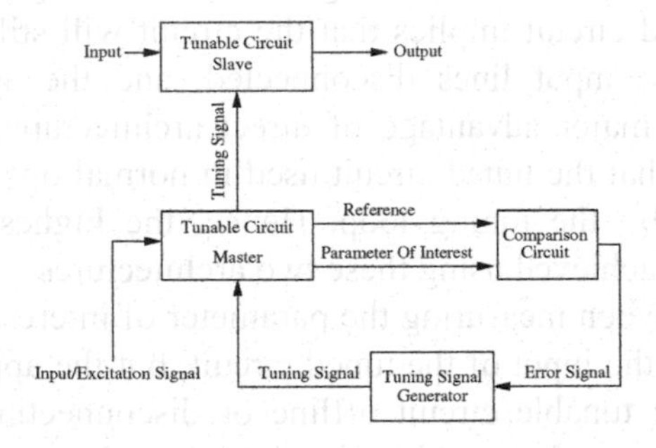

Figure 2-2. Indirect tuning architecture

The previous categorization of tuning techniques was based on how often tuning is performed, i.e. continuous or discontinuous, and whether the tuning loop is using the same circuit targeted by the tuning or just a copy of it, i.e. direct or indirect tuning. However, tuning loops could be also categorized based on the comparison nature. The nature of the reference usually dictates the nature of the comparison circuit. As will be discussed later in details, references could be either a voltage or a current level (magnitude), or a frequency (phase) of an oscillation. Therefore, tuning loops could be either Magnitude Locked Loops (MLL), or Phase Locked Loops (PLL) based. In MLL based tuning, the reference is a constant voltage or current level, and information about the parameter of interest is in another voltage or current level. By comparing the magnitude of these two levels, an error signal can be generated and then manipulated to generate the control signal (the tuning signal). Therefore, comparison circuits shown in Figs. 2-1 and 2-2 are magnitude comparators. In the analog domain, MLL-based comparison is normally done using a differential amplifier that subtracts the reference from the measured parameter of interest and yields an analog error signal. In the digital domain, the comparison is done using a comparator that yields a 1 or 0 based on its input levels. MLL-based tuning is generally susceptible to DC offsets in comparison circuits, which cause a systematic error in the tuned parameter. Note that MLL-based tuning relies on the ability to measure the parameter of interest and convert this measurement to a current or a voltage level, which even though possible, is not a straight forward process in some cases, especially if the parameter of interest is time-related (an oscillation frequency for example). PLL-based tuning on the other hand relies on a time reference instead of a level reference, i.e. a reference phase or frequency. In that case, the information about the measured parameter of interest is in the phase or frequency of a signal instead of its level. The phase of this signal is then compared to a reference phase using a phase detector in order to generate the error signal. In this case, comparison circuits shown in Figs. 2-1 and 2-2 are phase detectors.

Tuning architectures could also be categorized to analog-based or digital-based tuning. In analog-based tuning, the comparison process that generates the error signal and the further processing done on it in order to generate the tuning signal are purely done in the analog domain. Therefore, the tuning signal can assume infinite continuous values within the tuning range. In digital-based tuning, the measured parameter of interest and the reference signal are both converted to digital signals (either jointly or independently) using an Analog to Digital Converter (ADC). All comparisons and further processing is then performed in the digital domain, and the resulted digital tuning signal is converted to an analog signal using a Digital to Analog Converter (DAC), where it's then used to control the parameter of interest. In that case, the parameter of interest can only have finite discrete values in the tuning range, thus, the accuracy of the tuning is a function of the ADC and DAC resolution. Figure 2-3a shows a conceptual block diagram of a digital-based tuning architecture. In some cases though, if the parameter of interest can be controlled directly with a digital signal, i.e. the parameter of interest is digitally programmable, then the DAC can be eliminated altogether from the loop. An example of that is shown in Fig. 2-3b, where the parameter of interest is a function of capacitance. In this case, a bank of capacitors is used with individual capacitors directly turned on or off using digital signals.

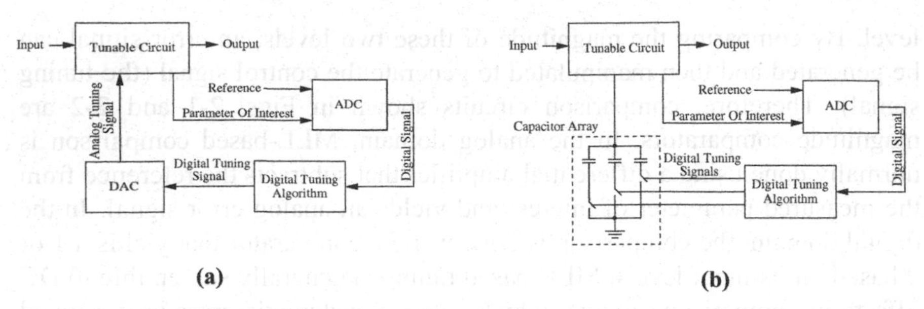

Figure 2-3. Digital based tuning: (a) with a DAC, and (b) with direct digital control.

Generally as shown by Figs. 2-1, 2-2 and 2-3, there is an absolute necessity for references and electronically tunable elements. In the next few sections, voltage and time references are discussed as well as electronically tunable elements including tunable resistors, capacitors, and transconductors.

2.1.1 Voltage, Current, and Time References

Unfortunately there are not many accurate absolute quantities that can be used in IC design. This fact is the main drive for developing circuits that solely depend on ratios between similar elements rather than absolute characteristics of each individual element. While this technique is very successful, still sometimes there is no way around depending on an absolute quantity. Generally, any quantity that has dimensions or units is always a suspect of being inaccurate unless it is equal to a universal constant that does not depend on the fabrication process, supply voltage, or temperature. Accurate references usually used in IC designs are voltage, current, and time references. Voltage references can usually be generated on-chip without requiring any external elements, while current references can be generated using an on-chip voltage reference in conjunction with an external resistor. Time references are usually generated using an external crystal oscillator in conjunction with an on-chip PLL (Phase Locked Loop).

2.1.1.1 Voltage References

A good voltage reference has to be insensitive to supply voltage, fabrication process, and temperature variations. There are numerous ways available in literature to generate a reference voltage[2]. This section focuses on circuits that generate the most accurate voltage reference, i.e. band-gap referenced circuits. The voltage drop across a forward-biased diode can be represented by the following equation[3]:

$$V_d = nV_T \ln\frac{I}{I_s} \tag{2-1}$$

where n is the emission constant (varies from 1 to 2), V_T is the thermal voltage, and I_s is the leakage current (typically around 10^{-15} A). The thermal voltage is represented by:

$$V_T = \frac{KT}{q} \tag{2-2}$$

where K is Boltzmann's constant (1.38 × 10^{-23} J/K), T is temperature in Kelvin, and q is the electrons charge (1.6 × 10^{-19} C). The thermal voltage V_T is equal to 26 mV at room temperature (T = 300 °K) and varies by 0.085 mV/°C with temperature. Even though it might look from Eq. 2-1 that V_d should have a positive temperature coefficient due to its proportionality with the thermal voltage, but it actually has a negative temperature coefficient. The reason behind that is that the leakage current I_s is a strong function of the band-gap energy of silicon E_g (eV) which can be represented by[3]:

$$I_s \propto e^{-E_g/KT} \tag{2-3}$$

$$E_g = 1.16 - \left(702\times10^{-6}\right)\left(\frac{T^2}{T+1108}\right) \tag{2-4}$$

Since the band-gap energy E_g of silicon has a negative temperature coefficient as shown by Eq. 2-4, then according to Eq. 2-3, the leakage current I_s will have a positive temperature coefficient. Furthermore, Eq. 2-1 shows that V_d is more sensitive to variations in I_s than to variations in V_T. Thus, V_d tends to have an effective negative temperature coefficient[2]. Typically, this temperature coefficient is around -2 mV/°C, which makes the forward-bias voltage of diodes not suitable as a temperature-stable reference.

One way to compensate for the negative temperature coefficient of the forward-bias voltage of diodes is to add to it another voltage level that has the same temperature coefficient but with a positive sign. In other words, a voltage level that is "Proportional To Absolute Temperature" (PTAT) is needed. Eq. 2-1 suggests that if the leakage current I_s was eliminated from the equation, then the resulted voltage will be proportional to V_T, and consequently will be a PTAT quantity. A simple way to eliminate I_s from Eq. 2-1 is by pumping two identical currents into two diodes that have different base areas. By subtracting the resulting forward-bias voltages of the two diodes, a PTAT voltage level is generated. A weighted version of this voltage level could then be added to the forward-bias voltage of another diode in order to generate a temperature-stable reference voltage. Circuits that employ the above methodology to generate accurate reference voltages are referred to as band-gap references.

Diodes are naturally implementable in CMOS technologies since different types of diffusion layers are available. Band-gap circuits use vertical PNP BJT transistors formed by the P^+ implant (the emitter), the n-well (the base), and the p-type substrate (the collector) in conjunction with CMOS transistors to generate a temperature-stable reference voltage. Figure 2-4 shows a simple band-gap circuit[2]. In this circuit, diode-connected BJT transistors D_2 and D_3 have N times the area of D_1, while the resistor in series with D_3 is L times the resistor in series with D_2. The CMOS current mirrors ensure that currents in D_1 and D_2 are identical, and therefore the current I in Fig. 2-4 could be written as:

$$I = \frac{nV_T \ln N}{R} \quad (2\text{-}5)$$

Note how the generated current is a PTAT quantity. The reference voltage can then be written as:

$$V_{ref} = ILR + V_{d3} \quad (2\text{-}6)$$

where V_{d3} is the forward-bias voltage of the diode-connected BJT D_3. Replacing I in Eq. 2-6 with Eq. 2-5, the reference voltage becomes:

$$V_{ref} = (Ln \ln N)V_T + V_{d3} \quad (2\text{-}7)$$

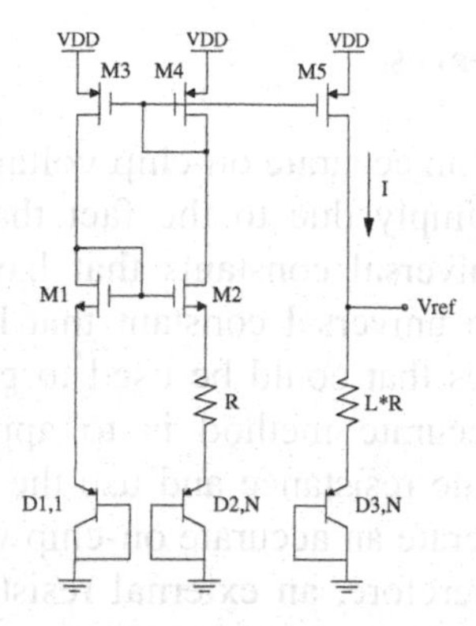

Figure 2-4. Schematic diagram of a simple band-gap circuit.

As previously mentioned, the forward-bias voltage of V_{d3} has negative temperature coefficient, while the voltage resulting across the resistors is a PTAT voltage, therefore both parts of Eq. 2-7 compensate for each other. The variation in the reference voltage V_{ref} with temperature is defined as:

$$\frac{\partial V_{ref}}{\partial T} = (Ln \ln N)\frac{\partial V_T}{\partial T} + \frac{\partial V_{d3}}{\partial T} \quad (2\text{-}8)$$

In order to have a temperature-stable reference voltage, Eq. 2-4 has to be equal to zero. The variation in the diode voltage with respect to temperature $\partial V_{d3} / \partial T$ is equal to -2 mV/°C, while the variation in the thermal voltage $\partial V_T / \partial T$ is 0.085 mV/°C at the room temperature[2] (T = 300 °K). Thus, in order for Eq. 2-4 to be equal to zero, the following equality has to hold:

$$Ln \ln N = \frac{2}{0.085} = 23.5 \quad (2\text{-}9)$$

By choosing the appropriate value for L, an accurate temperature-stable reference could be created. Since L is a ratio between two similar on-chip resistors, it is usually very accurate. Note also that CMOS transistors are only used as current mirrors and their absolute characteristics are of no concern. Band-gap references normally achieve an accuracy of ±3%. Fine tuning of L could be done through post fabrication trimming of the on-chip resistors to further improve the accuracy of the resulting voltage reference.

2.1.1.2 Current References

The ability to generate an accurate on-chip voltage reference without any external components is simply due to the fact that both the thermal and band-gap voltages are universal constants that have the units of voltage. Unfortunately, there is no universal constant that has the units of current. There are many techniques that could be used to generate accurate current references[2]. The most accurate method is to apply an accurate voltage reference across an accurate resistance and use the resulting current. While band-gap circuits can generate an accurate on-chip voltage, on-chip resistors can vary up to ±25%. Therefore, an external resistor has to be used along with a band-gap circuit and an op-amp to generate an accurate current reference as shown in Fig. 2-5. The op-amp copies the band-gap voltage reference across an accurate external resistor, and the current generated is then mirrored to be used in multiple locations in the IC. The obvious disadvantage of such a technique is the need of an external resistor.

2.1.1.3 Time References

Time references are not as easy to generate on-chip as in voltage references case. The reason behind that again is that there is no an on-chip universal constant that has the units of time. Even though there exist some techniques to generate on-chip accurate time constants based on RC products[4], but the most popular techniques heavily rely on an external crystal oscillator that is used in conjunction with a PLL to generate multiple frequency references. A piezoelectric crystal, such as quartz, exhibits electromechanical-resonance characteristics that are very stable with time and temperature and highly selective (having very high quality factors). The extremely stable resonance characteristics and the high quality factors of

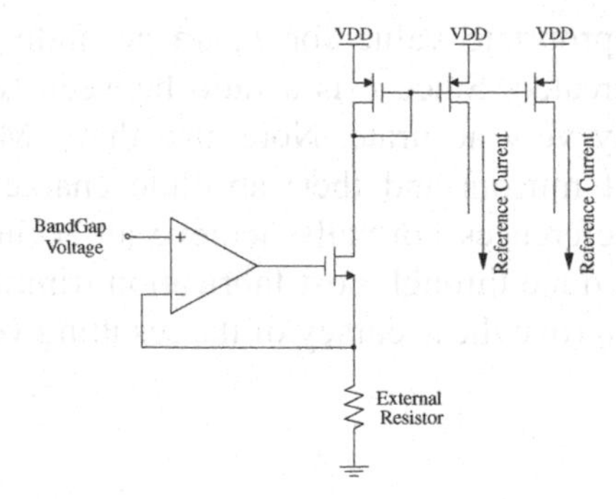

Figure 2-5. Reference current generator.

quartz crystals result in oscillators with very accurate and stable frequencies. Crystals are available with resonance frequencies in the range of few KHz to hundreds of MHz. Temperature coefficients of the frequency of oscillation ranges from 1 to 2 ppm per degree Celsius. However, crystal oscillators, being mechanical resonators, are fixed frequency circuits[5]. Therefore, there is always a need of an on-chip PLL to generate multiple frequencies, or multiple phases of the same frequencies. Those different frequencies and phases serve as accurate time references.

2.1.2 Electronically Tunable Elements

In order to tune for a certain parameter, it has to be electronically tunable using either a voltage, or current signal. Thus, the parameter of interest has to be a function of a variable resistance, capacitance, transconductance …etc. Those variable elements have to be either voltage or current controlled. In the next three sub-sections, examples of electronically tuned resistors, capacitors, and transconductors are presented.

2.1.2.1 Tunable Resistance

One of the most important electronically tunable parameters is the resistivity of the transistor itself. A MOS transistor current-voltage relationship can generally be represented by[6]:

$$I_D = (W/L) \int_{V_C=V_S}^{V_C=V_D} f(V_G, V_C)\, dV_C \tag{2-10}$$

with

$$f(V_G, V_C) = -\mu Q_C + \mu (KT/q)(dQ_C/dV_C) \tag{2-11}$$

where W and L are the physical dimensions of the transistor, V_S, V_G, and V_D are the source, gate, and drain voltages respectively, V_C is the channel voltage, μ is the current carriers mobility, KT/q is the thermal voltage, and Q_C is the channel charge. By first order approximation, Eqs. 2-10 and 2-11 are valid for all operation regions of the transistor including strong inversion, weak inversion, saturation, and non saturation regions[7]. The second part of Eq. 2-11 represents the weak inversion channel charge, while the first part represents the strong inversion channel charge. The function f can be a very complicated function in V_G and V_C if phenomena like mobility reduction and body effect are included. In order to simplify the computation without loosing much accuracy, the mobility is assumed to be constant ($\mu = \mu_0$) and that the transistor is operating in strong inversion. The channel charge could be represented by[8]:

$$Q_C = -[V_G - V_{TB}(V_C)]C_{ox} \tag{2-12}$$

where V_{TB} is the threshold voltage as a function of the channel voltage V_C, and C_{ox} is the gate oxide capacitance. Note that all voltages are referenced to the substrate potential. The variation in the threshold voltage with the channel potential, i.e. the body effect, can be approximated by the linear function[7]:

$$V_{TB}(V_C) = V_{TO} + \alpha V_C \tag{2-13}$$

where V_{TO} is the threshold voltage at $V_C = 0$, and α is a process dependent constant. In modern CMOS processes, α varies from 1.05 to 1.35. If the body effect is to be ignored, then α has to be assumed unity. Using Eqs. 2-12 and 2-13, Eq. 2-10 yields:

$$I_D = \mu_o C_{ox}(W/L) \int_{V_C=V_S}^{V_C=V_D} (V_G - V_{TO} - \alpha V_C)\, dV_C \tag{2-14}$$

The transistor is said to be in the triode region of operation as long as V_D is less than the channel saturation voltage $V_{Csat} = (V_G - V_{TO}) / \alpha$. In that case, by integrating Eq. 2-14, the drain current is represented by:

$$I_D = (K/2\alpha)\left[(V_G - V_{TO} - \alpha V_S)^2 - (V_G - V_{TO} - \alpha V_D)^2 \right] \tag{2-15}$$

where $K = \mu_o C_{ox} (W / L)$. When the drain voltage is higher than V_{Csat}, the channel pinches at a lower voltage than V_D and the transistor is said to be in the saturation region. In that case the drain current is represented by:

$$I_D = \mu_o C_{ox}(W/L) \int_{V_C=V_S}^{V_C=V_{Csat}} (V_G - V_{TO} - \alpha V_C)\, dV_C \tag{2-16}$$

$$I_D = (K/2\alpha)(V_G - V_{TO} - \alpha V_S)^2 \tag{2-17}$$

Note that the current equation in the saturation-mode of operation is simply the same as the triode-mode with the second part of Eq. 2-15 omitted.

As Eqs. 2-15 and 2-17 suggest, the current-voltage relationship of a transistor could be tuned using the gate voltage V_G, and therefore it can be used as a tunable resistance or a tunable voltage-to-current converter, i.e. transconductor. The only problem with that is the non linear behavior of the current-voltage relations presented in Eqs. 2-15 and 2-17. In order to use the transistor as a tunable resistor (or a transconductor), some nonlinearity cancellation techniques have to be employed. There are many nonlinearity cancellation techniques available in the literature based on both the triode and saturation-modes of the transistor's operation[1-7]. Generally, nonlinearity cancellation techniques rely on manipulating the terminal voltages and currents of single or multiple transistors to cancel out the nonlinear factors in Eqs. 2-15 and 2-17. The simplest nonlinearity cancellation technique is shown in Fig. 2-6a, in which a single transistor operating in the triode region is used with manipulated source and drain voltages to implement a tunable linear resistance. If the drain and source voltages are assumed to be:

$$V_D = V_B + V_X, \text{ and } V_S = V_B - V_X \tag{2-18}$$

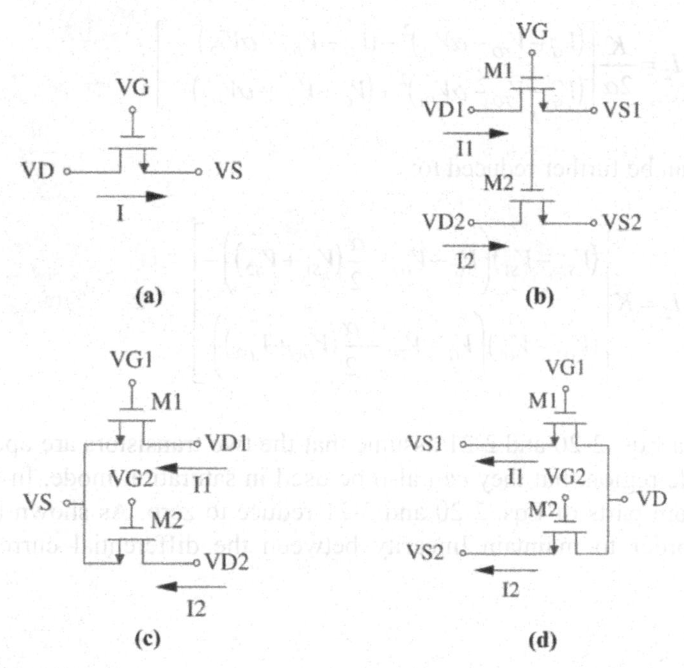

Figure 2-6. Linearization techniques based on: (a) a single transistor, (b) two transistors with common gates, (c) two transistors with common sources, and (d) two transistors with common drains.

where V_B is a constant voltage, then Eq. 2-15 reduces to:

$$R = \frac{2V_X}{I_D} = \frac{1}{K(V_G - V_{TO} - \alpha V_B)} \tag{2-19}$$

The resulted resistance is a strong function of temperature and process due to the appearance of K, V_{TO}, and α in Eq. 2-19, however, while those factors affect the absolute value of the resistance, they have not effect on its linearity. Inaccuracies in the absolute value can be easily tuned out by manipulating V_G.

Using a single transistor and manipulating terminal voltages to cancel the nonlinearities in the current-voltage relationship does not usually yield enough degrees of freedom on using the terminal voltages. In fact, as shown in the previous example, the drain and source voltages have to have the same common-mode, which is not always possible. Using multiple transistors to cancel the nonlinearity allows more degrees of freedom to the terminal voltages. Figure 2-6b shows two matched transistors sharing the same gate voltage V_G, while the drain and source voltages are separated. Using Eq. 2-15, the difference between the currents flowing into M_1 and M_2 is:

$$I_1 - I_2 = \frac{K}{2\alpha}\begin{bmatrix}(V_G - V_{TO} - \alpha V_{S1})^2 - (V_G - V_{TO} - \alpha V_{S2})^2 - \\ (V_G - V_{TO} - \alpha V_{D1})^2 + (V_G - V_{TO} - \alpha V_{D2})^2\end{bmatrix} \tag{2-20}$$

which can be further reduced to:

$$I_1 - I_2 = K\begin{bmatrix}(V_{S2} - V_{S1})\left(V_G - V_{TO} - \frac{\alpha}{2}(V_{S1} + V_{S2})\right) - \\ (V_{D2} - V_{D1})\left(V_G - V_{TO} - \frac{\alpha}{2}(V_{D1} + V_{D2})\right)\end{bmatrix} \tag{2-21}$$

Note that Eqs. 2-20 and 2-21 assume that the two transistors are operating in the triode region, but they can also be used in saturation-mode. In this case, the bottom parts of Eqs. 2-20 and 2-21 reduce to zero. As shown by Eq. 2-21, in order to maintain linearity between the differential current and a terminal voltage, there are two options: either eliminating the bottom part of Eq. 2-21 by forcing the drain voltages of both transistors to be equal, i.e. $V_{D1} = V_{D2} = V_D$, or eliminating the top part of Eq. 2-21 by forcing the source voltages of both transistors to be equal, i.e. $V_{S1} = V_{S2} = V_S$. In the first case, Eq. 2-21 reduces to:

$$I_1 - I_2 = K\left[(V_{S2} - V_{S1})\left(V_G - V_{TO} - \frac{\alpha}{2}(V_{S1} + V_{S2})\right)\right] \tag{2-22}$$

Furthermore, if the source voltages were manipulated such that $V_{S1} = V_B - V_X$ and $V_{S2} = V_B + V_X$, then Eq. 2-22 becomes:

$$R = \frac{2V_X}{I_1 - I_2} = \frac{1}{K(V_G - V_{TO} - \alpha V_B)} \tag{2-23}$$

which is again similar to Eq. 2-19. However, this method has the advantage of being valid for both the triode and saturation-modes of the two transistors since the bottom part of Eq. 2-21 that was eliminated by forcing the drain voltages of both transistors to be equal reduces to zero anyway in saturation-mode, which makes Eqs. 2-22 and 2-23 valid in that mode as well. If the source voltages are forced to be equal, Eq. 2-21 reduces to:

$$I_1 - I_2 = K\left[(V_{D1} - V_{D2})\left(V_G - V_{TO} - \frac{\alpha}{2}(V_{D1} + V_{D2})\right)\right] \tag{2-24}$$

Moreover, if the drain voltages were manipulated such that $V_{D1} = V_B + V_X$ and $V_{D2} = V_B - V_X$, then Eq. 2-22 becomes:

$$R = \frac{2V_X}{I_1 - I_2} = \frac{1}{K(V_G - V_{TO} - \alpha V_B)} \tag{2-25}$$

which is again similar to Eq. 2-19, but unlike Eq. 2-23, it is only valid when the two transistors are operating in the triode region.

Equations 2-22 and 2-24 were based on sharing the gate voltage between the two transistors in Fig. 2-6b. Another way of canceling nonlinearities is shown in Figs. 2-6c and 2-6d where the gate voltages are separated, while the source or the drain terminals are shared between the two transistors. If the two transistors are sharing the same source as in Fig. 2-6c, and assuming that both transistors are in triode-mode, the differential current becomes:

$$I_1 - I_2 = \frac{K}{2\alpha}\left[(V_{G1} - V_{G2})(V_{G1} + V_{G2} - 2V_{TO} - 2\alpha V_S)\right] - \frac{K}{2\alpha}\begin{bmatrix}(V_{G1} - V_{G2} - \alpha(V_{D1} - V_{D2}))\times \\ (V_{G1} + V_{G2} - 2V_{TO} - \alpha(V_{D1} + V_{D2}))\end{bmatrix} \tag{2-26}$$

and if drain voltages are equal, i.e. $V_{D1} = V_{D2} = V_D$, Eq.2-26 reduces to:

$$R = \frac{V_D - V_S}{I_1 - I_2} = \frac{1}{K(V_{G1} - V_{G2})} \tag{2-27}$$

The advantage of Eq. 2-27 over Eq. 2-19 is that the resistance value is not a function of the threshold voltage of the transistor, which results in a more accurate resistance. On the other hand, if the two transistors are sharing the same drain as shown in Fig. 2-6d instead of the same source like in Fig. 2-6c, and assuming triode-mode operation, the differential current becomes:

$$I_1 - I_2 = \frac{K}{2\alpha}\begin{bmatrix}(V_{G1} - V_{G2} - \alpha(V_{S1} - V_{S2}))\times \\ (V_{G1} + V_{G2} - 2V_{TO} - \alpha(V_{S1} + V_{S2}))\end{bmatrix} - \frac{K}{2\alpha}\left[(V_{G1} - V_{G2})(V_{G1} + V_{G2} - 2V_{TO} - 2\alpha V_D)\right] \tag{2-28}$$

and if the sources are forced to be at the same voltage, i.e. $V_{S1} = V_{S2} = V_S$, then Eq. 2-28 reduces to:

$$R = \frac{V_D - V_S}{I_1 - I_2} = \frac{1}{K(V_{G1} - V_{G2})} \tag{2-29}$$

which is the same resistance resulted in Eq. 2-27.

The structure in Fig. 2-6c, which results in a resistance value shown by Eq. 2-27 is only valid if both transistors are operating in the triode-mode. In order to expand the operation to the saturation-mode as well, the circuit is Fig. 2-7a could be used. In this case, the differential current could be represented by:

$$I_{O1} - I_{O2} = (I_1 - I_4) - (I_2 - I_3) \tag{2-30}$$

with

$$I_1 - I_4 = K\left[(V_{S34} - V_{S12})\left(V_{G1} - V_{TO} - \frac{\alpha}{2}(V_{S12} + V_{S34})\right)\right] \tag{2-31}$$

$$I_2 - I_3 = K\left[(V_{S34} - V_{S12})\left(V_{G2} - V_{TO} - \frac{\alpha}{2}(V_{S12} + V_{S34})\right)\right] \tag{2-32}$$

By substituting 2-31 and 2-32 in 2-30, the resulting resistance becomes:

$$R = \frac{V_{S34} - V_{12}}{I_{O1} - I_{O2}} = \frac{1}{K(V_{G1} - V_{G2})} \tag{2-33}$$

Since Eqs. 2-31 and 2-32 are valid in both triode and saturation-modes (given that all transistors have the same drain voltage), the resulted resistance is valid if transistor pairs M_1, M_4 and M_2, M_3 are both in saturation-mode, or both in triode-mode, or one pair in saturation-mode and the other in triode-mode. The circuit in Fig. 2-7a is also known as the four-quadrant multiplier, and it has the best linear behavior. In fact, it was shown that this circuit is practically insensitive to distributed effects in the transistor channel, and that it has inherent compensation to such effects[9,10].

A similar structure to the four-quadrant cell shown in Fig. 2-7a could be also used with the source terminals held at the same voltage for all transistors instead of the drain terminal. This structure is shown in Fig. 2-7b.

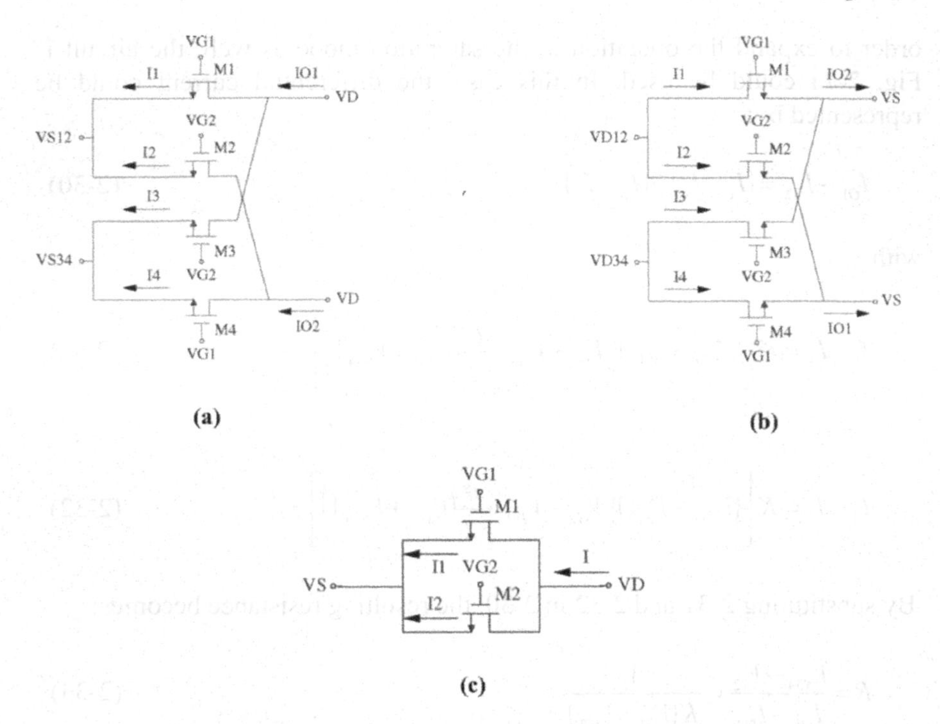

Figure 2-7. Linearization techniques based on: (a) the four-quadrant cell with common drains, (b) the four-quadrant cell with common sources, and (c) two transistors with common drains and common sources.

In this case, the differential current, assuming triode region operation could be represented by:

$$I_{O1} - I_{O2} = (I_1 - I_4) - (I_2 - I_3) \tag{2-34}$$

with

$$I_1 - I_4 = \frac{K}{2}\left[(V_{D12} - V_{D34})(V_{G1} - V_{TO} - \alpha(V_{D12} + V_{D34}))\right] \tag{2-35}$$

$$I_2 - I_3 = \frac{K}{2}\left[(V_{D12} - V_{D34})(V_{G2} - V_{TO} - \alpha(V_{D12} + V_{D34}))\right] \tag{2-36}$$

Substituting 2-35 and 2-36 in 2-34, the resulting resistance becomes:

$$R = \frac{V_{D12} - V_{D34}}{I_{O1} - I_{O2}} = \frac{1}{K(V_{G1} - V_{G2})} \tag{2-37}$$

Note that Eq. 2-37 is still only valid if all transistors are operating in the triode region. In fact, if all transistors are in the saturation region, the differential current becomes zero. Thus, the structure in Fig. 2-7a is more valuable than the one in Fig. 2-6b.

Another technique based on using the structure shown in Fig. 2-6d is shown in Fig. 2-7c, where both transistors share the same source and drain terminals and different gate voltages are applied. In this case, assuming triode-mode operation the total current can be represented by:

$$I = I_1 + I_2 \tag{2-38}$$

with

$$I_1 = K(V_D - V_S)\left(V_{G1} - V_{TO} - \frac{\alpha}{2}(V_D + V_S)\right) \tag{2-39}$$

$$I_2 = K(V_D - V_S)\left(V_{G2} - V_{TO} - \frac{\alpha}{2}(V_D + V_S)\right) \tag{2-40}$$

If the gate voltages are forced to be:

$$V_{G1} = V_C + V_D \text{, and } V_{G2} = V_C + V_S \tag{2-41}$$

where V_C is a control voltage, and by neglecting the body effect ($\alpha = 1$), Eq. 2-38 reduces to:

$$R = \frac{V_D - V_S}{I} = \frac{1}{2K(V_C - V_{TO})} \tag{2-42}$$

As discussed in this section, there are multiple ways to implement an electronically tunable resistance using linearization techniques of transistors in the triode-mode, or in the saturation-mode. While using transistors has the advantage of being electronically tunable and area efficient than passive resistors, yet it also suffers from major disadvantages. The first disadvantage is the nonlinear behavior of the resulting resistance. Even though the above nonlinearity cancellation techniques could be employed, they mostly rely on the classic square-law representation of transistors, which has not been very accurate in modeling nonlinearities in digital deep sub-micron CMOS processes. The second disadvantage is the that the usable voltage range of the resulted resistance is rather limited due to the fact that active transistors need a specific minimum voltage to be turned on, as well as any other range restriction to keep them in a specific operation mode (triode, or saturation). The third disadvantage is the poor noise performance of the resulted resistor since the transistors used are in direct contact with the substrate. Thus, their noise behavior is worse than a poly resistor for example.

A popular application of linearization techniques of transistors resistance is the implementation of voltage-controlled integrators that are used extensively in the area of MOSFET-C filters. In those filters, classical active RC filters are converted to all-MOS implementations with the aid of linearization techniques of MOSFET transistors[7]. Figure 2-8 shows an example of such an integrator using the four-quadrant structure shown in Fig. 2-7a. In this case, the integrator time-constant can be electronically tunable using V_{C1} and V_{C2}. In fact, using Eq. 2-33, the differential output voltage can be written as:

$$V_{O+} - V_{O-} = \frac{-K\left(V_{C1} - V_{C2}\right)}{C} \int_{-\infty}^{t} \left(V_{i+} - V_{i-}\right) dt \qquad (2\text{-}43)$$

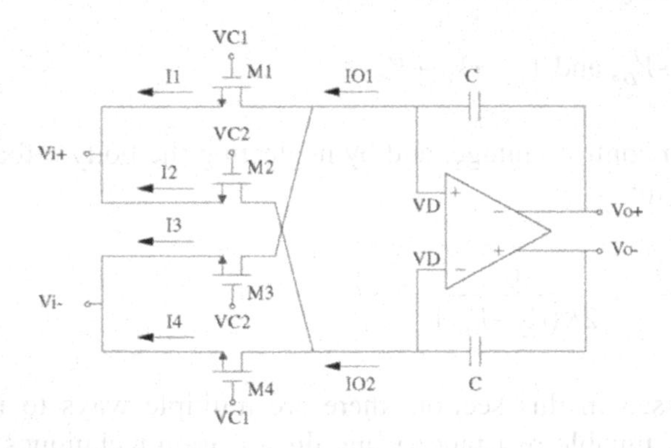

Figure 2-8. A tunable integrator.

2.1.2.2 Tunable Capacitors

Electronically tunable capacitors can greatly facilitate the tuning process of some specific blocks such as oscillators and filters. There are not as many ways to implement electronically tunable capacitors as in the case of resistors, and usually electronically tunable capacitors have a relatively limited tuning range (around 15%), which limits their usage in most cases to just fine tuning of an already course-tuned parameter. There are two main phenomena that are used to implement an electronically tunable capacitor (aka a varactor). The first phenomenon is the voltage-dependent capacitance behavior of a MOS capacitor biased in accumulation, flat-band, and depletion regions as will be discussed later in details in chapter 4. The second phenomenon is the voltage-dependent capacitance of depletion region in a lightly-doped reverse-biased PN junction. The unit depletion capacitance of a reverse-biased PN junction can be shown to be[3]:

$$C_{\mathrm{d}} = \frac{\varepsilon_{\mathrm{si}}}{W_{\mathrm{d}}} \tag{2-44}$$

where ε_{si} is the silicon dielectric constant, and W_d is the depletion region thickness. The depletion region thickness can also be shown to be[3]:

$$W_{\mathrm{d}} = \sqrt{\frac{2\varepsilon_{\mathrm{si}}\left(N_{\mathrm{a}} + N_{\mathrm{d}}\right)V_{\mathrm{d}}}{qN_{\mathrm{a}}N_{\mathrm{d}}}} \tag{2-45}$$

where V_d is the reverse voltage across the PN junction, N_a and N_d are the acceptor and donor doping density in the P and N sides respectively, and q is the electron charge. As shown by Eq. 2-45 the width of the depletion region can be controlled by the reverse voltage across the junction leading the capacitance in Eq. 2-45 to be electronically tunable.

MOS and PN junction capacitors, even though tunable, they have some serious limitations regarding their performance. They tend to have relatively poor noise performance due to their close proximity to the substrate. Additionally, once they are tuned, they are highly nonlinear with respect to voltage, which is a major source of distortion especially with large signals. Furthermore, they lack a degree of freedom since the control voltage that tunes the capacitance has to be one of the terminals of the capacitor itself, which substantially limits the configurations they can be used with since one of their terminals has to be always AC grounded.

2.1.2.3 Tunable Transconductors

A transconductor is simply a circuit block that converts a specific range of input voltage to a current signal with a linear transformation factor referred to as the transconductance G_m. This transconductance should not be confused with the transconductance of the transistor g_m. Transconductors can have a single-ended input and output, a differential input and a single-ended output, or a differential input and a differential output. Figure 2-9 shows a symbol for a fully differential transconductor. Transconductors have a wide range of applications in the area of analog signal processing. Those applications include active filters, amplifiers, equalizers, and numerous other applications[1-7]. Programmable transconductors are particularly important in the area of automatic tuning since they enable the tweaking of circuits behavior to compensate for process and temperature variations. Most programmable transconductors are controlled with a voltage signal that changes the transconductance value G_m. There have been numerous techniques in the literature for implementing those transconductors[11-14]. Generally, there are two types of transconductors. The first type is based on the triode-mode of operation of MOSFET transistors, while the second type is based on the saturation-mode of operation. Triode-mode based transconductors have better linear performance, i.e. less variations in G_m with respect to voltage, and they have better single-ended performance, i.e. each single-ended output current exhibits good linear behavior. On the other hand, transconductors based on the saturation-mode of operation of MOSFET transistors have the advantage of better speed performance, but they have moderate linear behavior when compared to triode based transconductors. Additionally, only the difference between the two single-ended output currents of the transconductor is linear, while each individual output suffers from significant nonlinearity[1].

In order to evaluate the performance of a transconductor, several factors have to be taken into consideration. The first factor is the input range in which the transconductor maintains a certain linear performance, i.e. output distortion levels. The second factor is the symmetry of the two differential outputs in fully differential transconductors, i.e. the quality of the fully differential nature of the transconductor. In other words, if the transconductance of the positive output is $G_{m+} = I_{O+}/V_{id}$ and of the negative output is $G_{m-} = I_{O-}/V_{id}$, then how close G_{m+} and G_{m-} are to each other, where I_{O+}, I_{O-}, and V_{id} are the positive output current, the negative output current, and the input differential voltage respectively. This is particularly important in determining common-mode distortion levels introduced at the outputs. The third factor is the control voltage range, i.e. the tuning range of the transconductor for a given input range and distortion levels at the output.

Figure 2-9 shows an example of a transconductor based on the triode-mode of operation of MOSFET transistors[15]. In this circuit, all transistors are operating in the saturation-mode except for M_9, which is operating in the triode-mode. Since the current in both M_1 and M_2 is constant, and assuming the two transistors are identical, then by using Eq. 2-17 the source voltages of these two transistors could be written as:

$$V_{S1} = \frac{V_{i+} - V_{TO}}{\alpha} - \sqrt{\frac{2I_1}{\alpha K}} \tag{2-46}$$

$$V_{S2} = \frac{V_{i-} - V_{TO}}{\alpha} - \sqrt{\frac{2I_1}{\alpha K}} \tag{2-47}$$

Since M_9 is operating in the triode-mode, then by substituting Eqs. 2-46, and 2-47 into 2-15, the current flowing in M_9 could be written as:

$$I_9 = 2\left(V_C - V_{CM} + \sqrt{\frac{2\alpha I_1}{K}}\right)V_{id} \tag{2-48}$$

where V_{id} is the differential input voltage $V_{i+} - V_{i-}$, and V_{CM} is the common-mode voltage $(V_{i+} + V_{i-})/2$. Since $I_{O+} = I_{O-} = I_9$, then the transconductance could be written as:

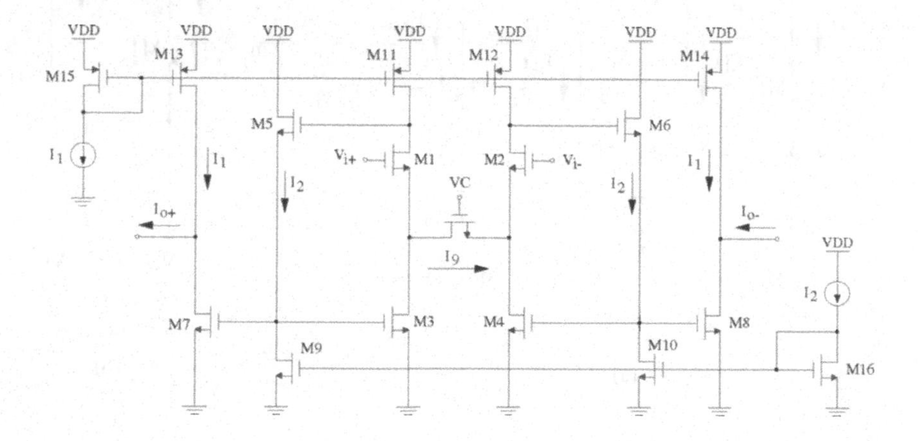

Figure 2-9. A triode-mode based transconductor.

$$G_m = \frac{I_{O+}}{V_{id}} = 2\left(V_C - V_{CM} + \sqrt{\frac{2\alpha I_1}{K}}\right) \tag{2-49}$$

As shown in Eq. 2-49, the transconductance is electronically tunable by V_C. However, the circuit in Fig. 2-9 suffers from some disadvantages. The main disadvantage is the distortion introduced at the outputs due to the presence of common-mode voltage in the transconductance formula. This is particularly a problem if the input signal has a time-variable common-mode. On the other hand, the circuit has the advantage of being capable of providing multiple output terminals by simply tapping off more current mirrors from M_3 and M_4, which could save a lot of die area and power consumption.

Figure 2-10a shows an example of a transconductor based on the saturation-mode MOSFET transistors[16,17,18]. The basic element of this transconductor is shown in Fig. 2-10b and is known as the COMFET structure, where both transistors are operating in the saturation-mode. For the COMFET structure, using Eq. 2-17 and assuming M_1 and M_2 are identical and ignoring the body effect ($\alpha = 1$), the differential current could be written as:

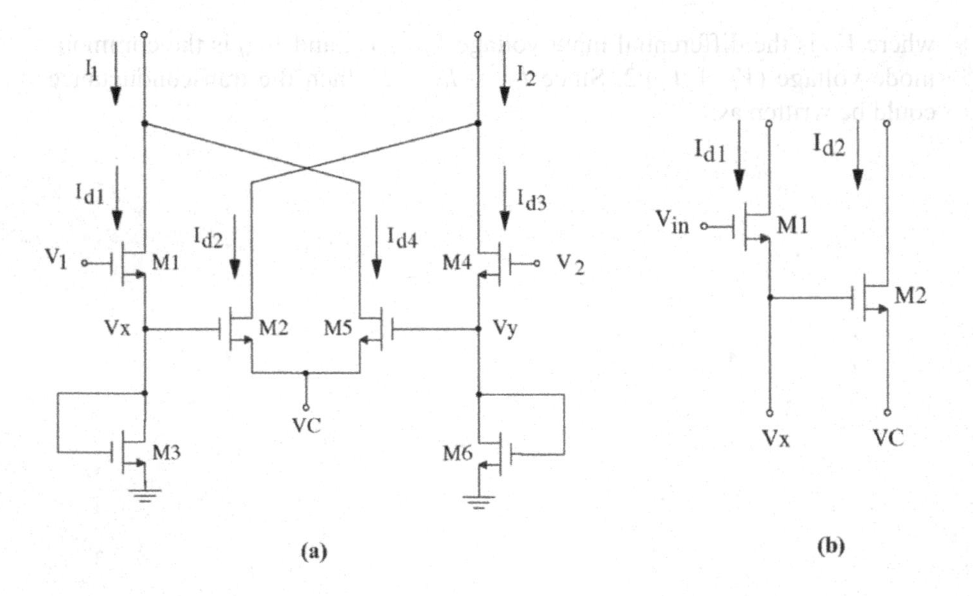

Figure 2-10. (a) saturation-mode based transconductor, (b) The COMFET structure.

$$I_{d1} - I_{d2} = (K/2)(V_{in} - 2V_X + V_C)(V_{in} - V_C - 2V_{TO}) \quad (2\text{-}50)$$

In the transconductor shown in Fig. 2-10a, and assuming that M_1, M_2, M_3, M_4, M_5, and M_6 are all identical, then V_X and V_Y could be written as:

$$V_X = \frac{V_1}{2} \text{ and } V_Y = \frac{V_2}{2} \quad (2\text{-}51)$$

Using Eqs. 2-50 and 2-51, the differential current in the transconductor, and the transconductance could be written as:

$$I_1 - I_2 = (I_{d1} - I_{d2}) - (I_{d3} - I_{d4}) = (K/2)(V_1 - V_2)V_C \quad (2\text{-}52)$$

$$G_m = \frac{I_1 - I_2}{V_1 - V_2} = (K/2)V_C \quad (2\text{-}53)$$

As shown in Eq. 2-53, the transconductance can be electronically tuned using V_C. However, the circuit in Fig. 2-10b suffers from some disadvantages. Like most saturation-based transconductors, only the differential current is linear with respect to the input voltage, while each individual output current is not. Additionally, the body effect of M_1, M_2, M_4, and M_5 adds to the distortion introduced to the output currents. Moreover, the input linear range is limited by conditions needed to turn on M_1 and M_3.

2.2 Post Fabrication Tuning

An alternative approach to automatic tuning is post fabrication tuning. It is based on tuning performance parameters after fabrication during the testing phase of the product. This method is particularly useful when the parameter of interest can not be automatically tuned on-chip. For instance, post fabrication tuning is very common for fine tuning band-gap voltage references since it is not possible to automatically tune them on-chip. The post fabrication tuning is based on developing an on-chip programmable block that can be controlled with a digital word. For example, if the parameter of interest is a function of a resistance, then this resistance could be implemented using multiple resistors in parallel with each resistor controlled by a switch that could be either turned on or off using a digital control signal. By successively changing the control word and comparing the parameter of interest to its targeted value, the control word that yields the best accuracy could be reached. In post fabrication tuning, the reference level is provided externally to the chip. Additionally, measuring the parameter of interest and comparing it to that reference level in order to generate the error signal are all performed externally rather than on-chip, where a software program on the production tester determines the control word accordingly. This control word could then be permanently hardwired on the chip using laser trimming, poly fuses, or an on-chip EPROM.

Laser trimming is based on using a laser beam to cut an on-chip metal line. By leaving or cutting the metal line, a digital 1 or a 0 could be permanently programmed inside the chip. Poly fuses on the other hand are just poly lines that could be blown electrically using a current signal, again a digital 1 or a 0 could be permanently programmed inside the chip depending on whether the fuse is blown or not. The EPROM is an electrically programmable read only memory implemented on-chip, a 1 or a 0 could be permanently written to that memory. It is worth mentioning that all the three previous methods could be employed only one time during the production test phase. Once the control word is programmed to the chip using one of the above methods, it can not be changed later and is permanently hardwired.

On-chip automatic tuning relies on an on-chip circuit that measures the error signal and sets the control signal accordingly. Thus, it is much cheaper than post fabrication tuning for three major reasons. First, it saves tester time leading to lower overall production cost. Second, it does not require any special processing steps as in laser trimming, poly fuses, or EPROM since the control signal is produced internally. Again a significant cost reduction. Third, on-chip tuning loops enable continuous tune-ability during the operation of the system, while post fabrication tuning using laser trimming, poly fuses, and EPROM offer only one time permanent programmability.

2.3 Summary

This chapter introduced the fundamental concepts of adaptive architectures including automatic and post fabrication tuning architectures. Automatic tuning architectures were categorized into two main categories: direct architectures where the tuned element is within the tuning loop, and indirect architectures where the tuned element is outside of the tuning loop. Both architectures could be continuous, in which the tuning loop is continuously running during normal operation of the system, or discontinuous, in which the tunable element is taken offline during the tuning process until a control signal value has been reached, then this value is held and the tunable circuit is put back online. The basic elements required in automatic tuning architectures were also presented including references (voltage, current, and time references) and tunable elements (resistors, capacitors, and transconductors).

LIST OF REFERENCES

1. David A. Johns, Ken Martin, "Analog Integrated Circuit Design," John Wiley & Sons, New York, 1997.
2. R. Jacob, Harry W. Li, David E. Boyce, "CMOS Circuit Design, Layout, and Simulation," IEEE Press Series on Microelectronic Systems, New York, 1998.
3. Yuan Taur, Tak H. Ning, "Fundamentals Of Modern VLSI Devices," Cambridge University Press, Cambridge 1998.
4. R. A. Blauschild, "An Integrated Time Reference," *ISSCC Dig. Tech. Papers*, PP. 56-57, Feb. 1994.
5. Adel Sedra, Kenneth Smith "Microelectronic Circuits," Third Edition, Saunders College Publishing, Toronto 1991.
6. K. Bult and G. J. G. M. Geelen, "An Inherently Linear and Compact MOST-Only Current-Division Technique," *IEEE Journal of Solid-State Circuits*, vol. 27, No. 6, PP. 1730-1735, Dec. 1992.
7. Mohammed Ismail, Terri Fiez, "Analog VLSI Signal and Information Processing," McGraw-Hill, New York, 1994.
8. H. Wallinga and K. Bult, "Design and Analysis of CMOS Analog Processing Circuits by Means of a Graphical MOST Model," *IEEE Journal of Solid-State Circuits*, vol. 24, No. 3, PP. 672-680, Jun. 1989.
9. N. I. Khachab and M. Ismail, "Linearization techniques for n^{th}-order sensor models in MOS VLSI technology," *IEEE Trans. Circuits. Syst.*, vol. 38,PP. 1439-1450, Dec. 1991.
10. M. Ismail and D. Rubin, "Improved circuits for the realization of MOSFET-capacitor filters," *IEEE Int. Symp. Circuits. Syst.*,PP. 1186-1189, May. 1986.
11. S. T. Dupuie and M. Ismail, "High frequency CMOS transconductors," *in Analog IC Design: the current-mode approach* (C. Toumazou, F.J. Lidgey, and D. G. Haigh, eds), ch. 5, London: Peter Peregrinus Ltd., 1990.
12. P. E. Allen and D. R. Holberg, "CMOS Analog Circuit Design," Holt, Rinehart and Winston, 1987.
13. M. Ismail, "Four-transistor continuous-time MOS transconductor," *Electronics Letters*, vol. 23, PP. 1099-1100, Sept. 1987.
14. P. Ryan and D. G. Haigh, "Novel fully-differential MOS transconductor for integrated continuous-time filters," *Electronics Letters*, vol. 23, PP. 742-743, Jul. 1987.
15. D. R. Welland, S. M. Phillip, Ka. Y. Leung, G. T. Tuttle, S. T. Dupuie, D. R. Holberg, R. V. Jack, N. S. Sooch, K. D. Anderson, A. J. Armstrong, R. T. Behrens, W. G. Bliss, T. O. Dudley, W. R. Foland, N. Glover, L. D. King, "A digital read/write channel with EEPR4 detection," *ISSCC Dig. Tech. Papers*, PP. 276-277, Feb. 1994.
16. M. C. H. Cheng and C. Toumazou, "Linear composite MOSFETS (COMFET)," *Electronics Letters*, PP. 1802-1804, Sept. 1991.
17. E. Seevinck and R. F. Wassenaar, "A versatile CMOS linear transconductor/square-law function circuit," *IEEE Journal of Solid-State Circuits*, vol. SC-22, PP. 366-377, Jun. 1987.
18. S. C. Huang and M. Ismail, "Linear tunable COMFET transconductors," *Electronics Letters*, vol. 29, PP. 459-461, Mar. 1993.

Chapter 3

TUNABLE ELEMENTS

Tunable elements are essential for any adaptive architecture to operate successfully. By definition, a tunable circuit is a circuit that enables the change of one or more of its key parameters using a voltage or a current control signal. Thus, the presence of such circuits provides a knob to tweak the performance of the whole system, and if implemented within an adaptive loop, then this tweaking could be achieved automatically as outlined in chapter 2. Examples of available tunable elements would be tunable resistors, capacitors, transconductors, amplifiers, and offset comparators. Chapter 2 gave an overview of the techniques available to implement tunable resistors, capacitors, and transconductors. Since voltage-controlled transconductors are particularly very popular, this chapter focuses on discussing circuit implementations that enable higher performance, wider control range, and more robust programmable performance. In addition to voltage-controlled transconductors, and since programmable comparators were not discussed in chapter 2, this chapter will also discuss the general characteristics of comparators with an emphasis on available techniques for implementing a specific class of them, i.e. Offset Comparators. It then proceeds with presenting a circuit design technique for implementing high-speed voltage-controlled offset comparators, where higher performance and more accurate control over the offset could be achieved in a robust fashion.

3.1 Voltage-Controlled Transconductors

Transconductors have a wide range of applications in the area of analog signal processing, such as active filters, amplifiers, equalizers, and numerous other applications[1,2]. Fully differential, programmable transconductors are particularly very important class of transconductors especially in the area of automatic tuning and adaptive architectures in general since they enable tweaking of system behavior to compensate for process and temperature variations. In order to evaluate the performance of a fully differential transconductor, several factors have to be taken into consideration. The first factor is the input range for which a certain linear performance is maintained, i.e. constant G_m, which is very important for determining the level of distortion in the output currents for a given input voltage range. The second factor is the symmetry of the two differential outputs, i.e. the quality of the differential nature of the transconductor. In other words, if the transconductance of the positive output is $G_{m+}=I_{O+}/V_{id}$ and for the negative output is $G_{m-}=I_{O-}/V_{id}$, then how close G_{m+} and G_{m-} are to each other, where I_{O+}, I_{O-}, and V_{id} are the positive output current, negative output current, and input differential voltage respectively. This is particularly important for determining the level of common-mode distortion the transconductor introduces at the outputs. The third factor is the control voltage range, i.e. the tuning range of the transconductor given a certain input voltage range and distortion levels at the outputs. This factor is particularly important at low supply voltages. The performance factors described above are especially difficult to maintain in digital CMOS technologies, particularly if digital core devices are used with channel lengths close to the feature. This is mainly because core transistors are optimized for maximum digital switching speed, and therefore have poor analog performance. In the next 3 sections, a low-voltage, highly-linear, voltage-controlled transconductor with wide input/control voltage range is presented. The transconductor is suitable for implementation using digital core transistors with channel lengths of no more than double the feature size, where a simple technique is used to cancel second-order nonlinearities introduced by MOS transistors in triode-mode of operation. Significant improvement in linearity, input range, control voltage range, and differential nature of the transconductor is reported. A simple triode-based programmable transconductor is used as a starting point for the development of the improved implementation, and a comparison between the two transconductors is discussed to highlight the advantages of the improved technique.

3.1.1 A Triode-Based Voltage-Controlled Transconductor

As mentioned in chapter 2, a transconductor is simply a circuit block that converts an input voltage to a current signal with a linear transformation factor referred to as the transconductance G_m. The input voltage range that maintains a specific linear performance i.e. constant G_m, the symmetry of the two differential outputs, i.e. quality of the transcoductor's differential nature are important performance parameters. Additionally, the control voltage range, i.e. the tuning range of the transconductor for a given input range and

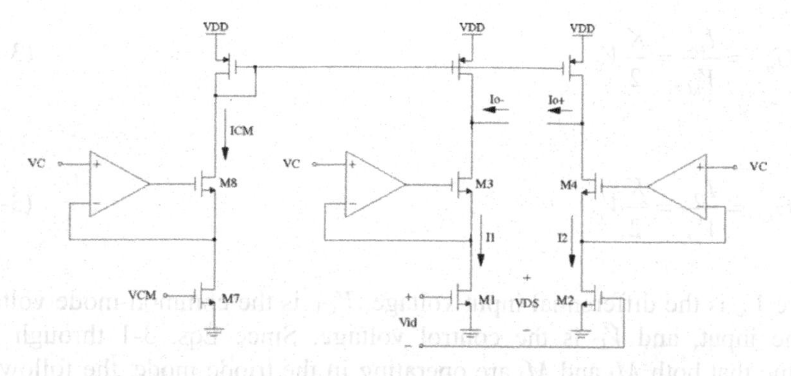

Figure 3-1. Schematic diagram of a linear mode based transconductor.

distortion levels at the output is also an important factor that determines the overall quality of the transconductor. Figure 3-1 shows a transconductor that is based on the triode-mode of operation of the MOSFET transistor[2]. As explained in chapter 2, triode-mode-based transconductors have better linear performance than saturation-mode-based transconductors. For the circuit shown in Fig. 3-1 and assuming the ideal triode-mode current equation[2]:

$$I_1 = K\left[\left(V_{CM} + \frac{V_{id}}{2} - V_T\right)V_C - \frac{V_C^{\,2}}{2}\right] \tag{3-1}$$

$$I_2 = K\left[\left(V_{CM} - \frac{V_{id}}{2} - V_T\right)V_C - \frac{V_C^{\,2}}{2}\right] \tag{3-2}$$

$$I_{CM} = K\left[\left(V_{CM} - V_T\right)V_C - \frac{V_C^{\,2}}{2}\right] \tag{3-3}$$

$$I_{O+} = I_{CM} - I_2 = \frac{K}{2}V_{id}V_C \tag{3-4}$$

$$I_{O-} = I_1 - I_{CM} = \frac{K}{2}V_{id}V_C \tag{3-5}$$

$$G_{m+} = \frac{I_{O+}}{V_{id}} = \frac{K}{2} V_C \tag{3-6}$$

$$G_{m-} = \frac{I_{O-}}{V_{id}} = \frac{K}{2} V_C \tag{3-7}$$

where V_{id} is the differential input voltage, V_{CM} is the common-mode voltage of the input, and V_C is the control voltage. Since Eqs. 3-1 through 3-7 assume that both M_1 and M_2 are operating in the triode-mode, the following condition has to be satisfied:

$$V_C < V_{CM} - \frac{V_{id}}{2} - V_T \tag{3-8}$$

Equations 3-1 through 3-7 suggest that the transconductor is achieving the desired function in a perfectly linear fashion. However, in order to evaluate the linear performance of the transconductor, a more accurate model of the triode-mode operation of the transistor that models second-order effects has to be used. Equations 3-1 and 3-2 are assuming that in the triode-mode of operation the current is linearly proportional to the gate-source voltage (the ideal MOS current formula), and that is not very accurate especially for short-channel devices. In order to model the nonlinearity with respect to the gate-source voltage of a MOSFET, the current equation of the triode-mode of operation could be rewritten as:

$$I_D = K\left[\left(\left(V_{GS} + \beta_1 V_{GS}^2 + \beta_2 V_{GS}^3 + \dots\dots\dots\right) - V_T\right)V_{DS} - \frac{V_{DS}^2}{2}\right] \tag{3-9}$$

where β is a process-dependent parameter. Since the cubic and higher-order terms are very small, they can be ignored, and only the square-term is used. Equation 3-8 will then be reduced to:

$$I_D = K\left[\left(\left(V_{GS} + \beta_1 V_{GS}^2\right) - V_T\right)V_{DS} - \frac{V_{DS}^2}{2}\right] \tag{3-10}$$

Using Eq. 3-10, Eqs. 3-1 through 3-7 for the circuit in Fig. 3-1 can be rewritten as:

$$I_1 = K\left[\left(V_{CM} + \frac{V_{id}}{2}\right) + \beta_1\left(V_{CM} + \frac{V_{id}}{2}\right)^2 - V_T\right]V_C - \frac{V_C^2}{2}\right] \quad (3\text{-}11)$$

$$I_2 = K\left[\left(V_{CM} - \frac{V_{id}}{2}\right) + \beta_1\left(V_{CM} - \frac{V_{id}}{2}\right)^2 - V_T\right]V_C - \frac{V_C^2}{2}\right] \quad (3\text{-}12)$$

$$I_{CM} = K\left[\left(V_{CM} + \beta_1 V_{CM}^2 - V_T\right)V_C - \frac{V_C^2}{2}\right] \quad (3\text{-}13)$$

$$I_{O+} = I_{CM} - I_2 = K\left[\left(\frac{V_{id}}{2} + \beta_1\left(2V_{CM} - \frac{V_{id}}{2}\right)\frac{V_{id}}{2}\right)V_C\right] \quad (3\text{-}14)$$

$$I_{O-} = I_1 - I_{CM} = K\left[\left(\frac{V_{id}}{2} + \beta_1\left(2V_{CM} + \frac{V_{id}}{2}\right)\frac{V_{id}}{2}\right)V_C\right] \quad (3\text{-}15)$$

$$G_{m+} = \frac{\partial}{\partial V_{id}} I_{O+} = K\left[\frac{1}{2} + 2\beta_1 V_{CM} - \frac{\beta_1}{4}V_{id}\right]V_C \quad (3\text{-}16)$$

$$G_{m-} = \frac{\partial}{\partial V_{id}} I_{O-} = K\left[\frac{1}{2} + 2\beta_1 V_{CM} + \frac{\beta_1}{4}V_{id}\right]V_C \quad (3\text{-}17)$$

Equations 3-10 through 3-17 could now be used to provide a more accurate assessment of performance limitations in the circuit. The first performance limitation in the circuit shown in Fig. 3-1 is the linearity of the output currents with respect to the differential input voltage. As Eqs. 3-16 and 3-17 show, the value of both G_{m+} and G_{m-} is a function of the differential input voltage V_{id}, which introduces nonlinearity to the output currents and consequently increases the total harmonic distortion in the output signal. Hence, if a specific linear performance is required, the input voltage range has to be limited to a small value. This fact further decreases the valuable input voltage range beyond the condition specified in Eq. 3-8. The second limitation is that G_{m+} and G_{m-}, are not equal, which means that the circuit is not truly fully differential, and will introduce common-mode distortion at the output. The third limitation is the control voltage range of the circuit. Since the circuit in Fig. 3-1 assumes that M_1 and M_2 are operating in the triode-mode of operation, the condition shown in Eq. 3-8 has to be satisfied. Hence, if a wide input range is needed, the control voltage range has to decrease and vice versa, which is a disadvantage at low supply voltages. In some cases though, even with losing linearity due to exceeding the control voltage limit shown in Eq. 3-8, it is desirable to still be able to control the transconductance beyond the limit shown in Eq. 3-8. In fact, in order to evaluate the effect on the transconductance if the condition stated in 3-8 is not met, and assuming positive differential voltage, the current equation for M_2 has to be changed to the saturation-mode of operation formula:

$$I_2 = \frac{K}{2}\left(V_{CM} - \frac{V_{id}}{2} - V_T\right)^2 \tag{3-18}$$

hence Eqs. 3-16 and 3-17 will be changed to:

$$G_{m+} = \frac{\partial}{\partial V_{id}} I_{O+} = K\left(V_{CM} - \frac{V_{id}}{2} - V_T\right) \tag{3-19}$$

$$G_{m-} = \frac{\partial}{\partial V_{id}} I_{O-} = K\left(\frac{1}{2} + 2\beta_1 V_{CM} + \frac{\beta_1 V_{id}}{4} - V_T\right) Vc \tag{3-20}$$

As Eqs. 3-19 and 3-20 show, the transconductance of the positive output saturates and no longer depends on V_C, which means it can't be controlled any further, while the transconductance of the negative output, even though suffers from nonlinearity because of the presence of V_{id}, but it can still be controlled further with V_C. Since the transconductance of one output saturates while the transconductance of the other output does not when condition 3-8 is not met, the circuit completely losses its fully differential nature. This is in addition to the fact that distortion will be introduced due to the dependence of G_{m-} on V_{id} as shown in Eq. 3-20. The fourth limitation in the circuit in Fig. 3-1 is that it needs a third op-amp to generate the I_{CM} signal. This is in addition to the extra circuitry needed to extract the common-mode voltage of the input signal, which increases the size and power consumption of the transconductor. In the next section, a modification to the circuit shown in Fig. 3-1 will be presented. The introduced modification extends the control voltage range and also significantly improves the linearity of the circuit. The modification also makes the transconductor truly fully differential.

3.1.2 An Improved Voltage-Controlled Transconductor

The modification to the circuit shown in Fig. 3-1 significantly improves the linearity of the transconductor by making both G_{m+} and G_{m-} independent from the differential input signal[3,4]. The modification also makes the design truly fully differential by making G_{m+} and G_{m-} equal for all input and control voltage ranges. The circuit also significantly increases the control voltage range by making both G_{m+} and G_{m-} proportional to V_C even when transistors M_1 and M_2 enter the saturation region. The circuit also eliminates the need for a third op-amp or a common-mode extraction circuit, which will save power and area. The transconductor is shown in Fig. 3-2. The positive input stage is comprised of M_1, M_5, and M_9, while the negative input stage is comprised of M_2, M_6, and M_{10}. Both the positive and negative input stages are identical. M_3, M_7, and M_{11} are identical to M_1, M_5, and M_9 respectively, while M_4, M_8, and M_{12} are identical to M_2, M_6, and M_{10} respectively. Since transistors M_1 and M_5 are connected in a cascode configuration as well as transistors M_3 and M_7, and assuming that M_5 and M_7 are operating in the saturation-mode (a condition that could be met by properly sizing M_9, M_{10}, M_{11}, and M_{12}), the output impedance at the drains of both M_5 and M_7 will be substantially high. This means that the drain voltages of M_5 and M_7 will have a small effect on the values of the currents I_3 and I_4 respectively. Since M_5 and M_7 are identical and they have the same gate voltage, then the currents I_3 and I_4 will be identical if the source voltages of M_5 and M_7 are identical. Furthermore, since M_1 and M_3 are identical and they have the same gate to source voltage V_{GS}, and since they both operate in the triode-mode, then the currents I_3 and I_4 will be identical if the drain voltages of M_1 and M_3 are identical. If we assume that I_4 is less than I_3, then the source voltage of M_7 (the drain of M_3) will have to be higher than the source voltage of M_5 (the drain of M_1), but having a higher drain voltage on M_3 means that I_4 will be higher than I_3, which contradicts the assumption. Furthermore, if we assume that I_3 is less than I_4, then the source voltage of M_5 (the drain of M_1) will have to be higher than the source voltage of M_7 (the drain of M_3), but having a higher drain voltage on M_1 means that I_3 will be higher than I_4, which again contradicts the assumption. Considering the arguments mentioned above, the only stable condition for the circuit will be that the currents I_3 and I_4 flowing in transistors M_1 and M_3 respectively are essentially equal, and that the source voltages of M_5 and M_7 (the drain voltages of M_1 and M_3 respectively)

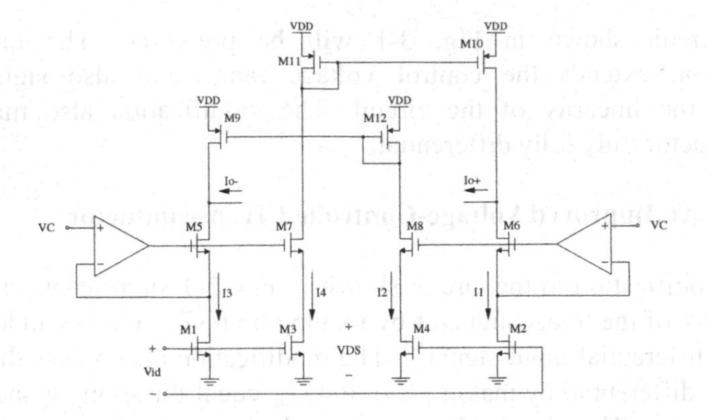

Figure 3-2. Schematic diagram of the improved transconductor.

are also essentially identical. In other words, a virtual short circuit between the drains of transistors M_1 and M_3 is established. Since the op-amp sets the drain of M_1 to V_C, the drain of M_3 will also be set to V_C. This guarantees that the control voltage V_C is controlling the drains of both M_1 and M_3, and that the currents I_3 and I_4 are identical without any need for an extra op-amp to force V_C on the drain of transistor M_3. The same analysis could be applied to the negative input stage as well.

In order to demonstrate how the circuit improves output linearity as well as input range of the transconductor, using Eq. 3-9, the following set of equations are derived:

$$I_1 = I_2 = K\left[\left(\left(V_{CM} - \frac{V_{id}}{2}\right) + \beta_1\left(V_{CM} - \frac{V_{id}}{2}\right)^2 - V_T\right)V_C - \frac{V_C^2}{2}\right] \quad (3\text{-}21)$$

$$I_3 = I_4 = K\left[\left(\left(V_{CM} + \frac{V_{id}}{2}\right) + \beta_1\left(V_{CM} + \frac{V_{id}}{2}\right)^2 - V_T\right)V_C - \frac{V_C^2}{2}\right] \quad (3\text{-}22)$$

$$I_{O+} = I_4 - I_1 = K\left[\left(1 + 2\beta_1 V_{CM}\right)V_{id}V_C\right] \quad (3\text{-}23)$$

$$I_{O-} = I_3 - I_2 = K\left[\left(1 + 2\beta_1 V_{CM}\right)V_{id}V_C\right] \quad (3\text{-}24)$$

$$G_{m+} = \frac{\partial}{\partial V_{id}} I_{O+} = K\left[\left(1 + 2\beta_1 V_{CM}\right) V_C\right] \tag{3-25}$$

$$G_{m-} = \frac{\partial}{\partial V_{id}} I_{O-} = K\left[\left(1 + 2\beta_1 V_{CM}\right) V_C\right] \tag{3-26}$$

Comparing Eqs. 3-23 through 3-26 to Eqs. 3-14 through 3-17, it is clear that the proposed modification forced both the positive and negative current outputs, as well as the transconductances to be equal. This equality makes the design truly fully differential and reduces common-mode distortion as well as offsets introduced when I_{O+} and I_{O-} are not equal. Equations 3-25 and 3-26 show that the transconductance of both current outputs of the circuit is also linear and constant with respect to the differential input voltage (as opposed to Eqs. 3-16 and 3-17). This improvement significantly reduces the total harmonic distortion at the output currents. Hence, if certain nonlinearity level in the transconductance is acceptable, the circuit in Fig. 3-2 guarantees a much wider differential input range than the circuit in Fig. 3-1 for the same nonlinearity level. In order to also see how the circuit extends the control voltage range as well, if M_2 and M_4 enters the saturation-mode of operation (Eq. 3-8 is not satisfied), while M_1 and M_3 stay in the linear region, then the following set of equations could be derived using Eqs. 3-11 and 3-18:

$$G_{m+} = \frac{\partial}{\partial V_{id}} I_{O+} = \frac{K}{2}\left[\left(V_{CM} - \frac{V_{id}}{2} - V_T\right)\right] + K\left[\frac{1}{2} + \beta_1\left(V_{CM} + \frac{V_{id}}{2}\right)\right] V_C \tag{3-27}$$

$$G_{m-} = \frac{\partial}{\partial V_{id}} I_{O-} = \frac{K}{2}\left[\left(V_{CM} - \frac{V_{id}}{2} - V_T\right)\right] + K\left[\frac{1}{2} + \beta_1\left(V_{CM} + \frac{V_{id}}{2}\right)\right] V_C \tag{3-28}$$

Comparing Eqs. 3-27 and 3-28 to Eqs. 3-19 and 3-20, Eqs. 3-27 and 3-28 show that for higher values of V_C when Eq. 3-8 is not satisfied and some of the transistors operate in the saturation-mode, the transconductances, although dependent on V_{id}, still can be controlled by V_C in a linear fashion. This is very beneficial especially at low supply voltages since it extends the range of control V_C have over the transconductance. Equations 3-27 and 3-28 also show that both G_{m+} and G_{m-} are equal, hence maintaining the truly fully differential nature of the circuit even when transistors enter into the saturation-mode of operation.

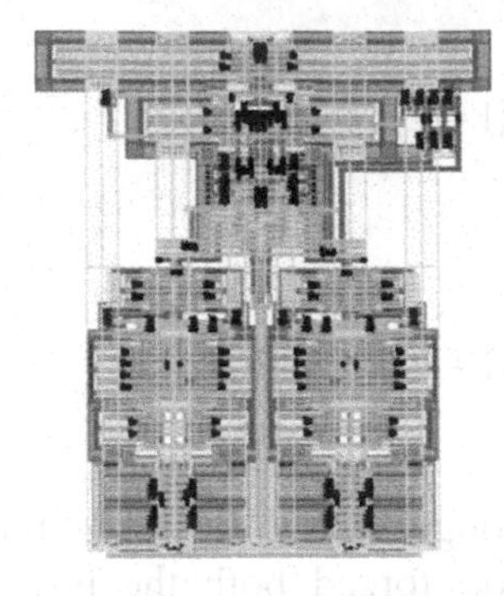

Figure 3-3. Layout of the transconductor shown in Fig. 3-2.

3.1.3 Simulations Results

In order to assess the performance of the circuit, both transconductors (Figs. 3-1 and 3-2) were implemented on a 180 nm digital CMOS process with a nominal supply voltage of 1.8V. For the transconductor in Fig. 3-2, the input transistors M_1, M_2, M_3, and M_4 had an aspect ratio of 0.3μm/0.3μm, while transistors M_5, M_6, M_7, and M_8 had an aspect ratio of 6μm/0.36μm. The mirror transistors M_9, M_{10}, M_{11}, and M_{12} had an aspect ratio of 16.5μm/0.36μm. Note that all the used transistors are digital core transistors with channel lengths of no more than double the feature size of the process. The two op-amps used in the design are implemented using a simple current mirror OTA with a PMOS input stage and a high impedance output stage. The DC gain of the op-amp is designed to be around 30 db with unity-gain frequency around 75 MHz. Figure 3-3 shows the layout of the transconductor in Fig. 3-2. The circuit occupies around 1945 μm^2, and consumes an average power of 418 μw at 65 MHz.

Simulation results show the improvements that the circuit in Fig. 3-2 has over the original circuit shown in Fig. 3-1. Figures 3-4 and 3-5 show the output currents of both transconductors shown in Fig. 3-1 and Fig. 3-2. Note the linearity difference in the currents. To emphasize the improvement in linearity, Figs. 3-6, 3-7, and 3-8 show the transconductances of both circuits at different control voltages, which ideally should be constant across the input voltage range for a perfectly linear transconductor. Those figures clearly show the significant improvement in the transconductance variation with input voltage, i.e. much less variation in the transconductance value with respect to the input differential voltage, an advantage that significantly reduces total harmonic distortion.

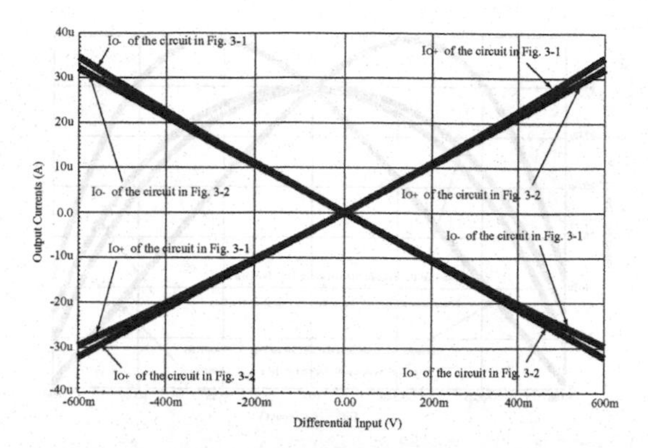

Figure 3-4. Output currents of transconductors in Fig. 3-1 and 3-2 at 100mV control voltage.

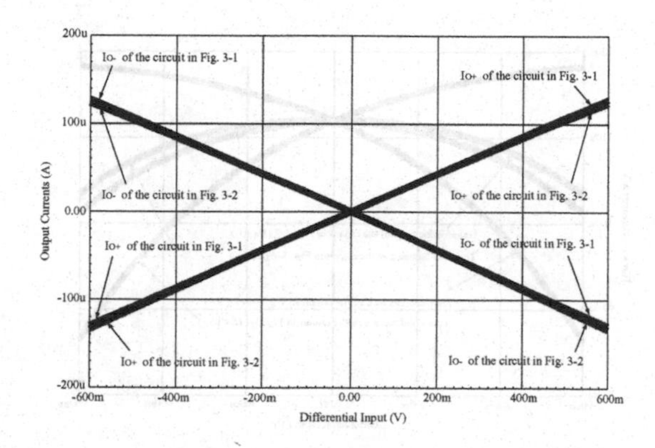

Figure 3-5. Output currents of transconductors in Fig. 3-1 and 3-2 at 1V control voltage.

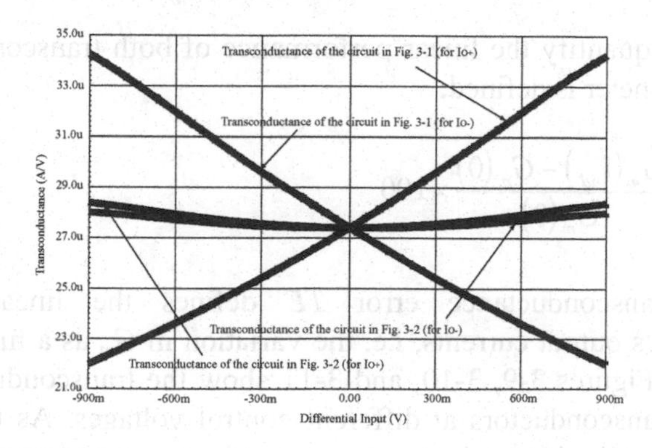

Figure 3-6. G_{m+} and G_{m-} of transconductors in Fig. 3-1 and 3-2 at 100mV control voltage.

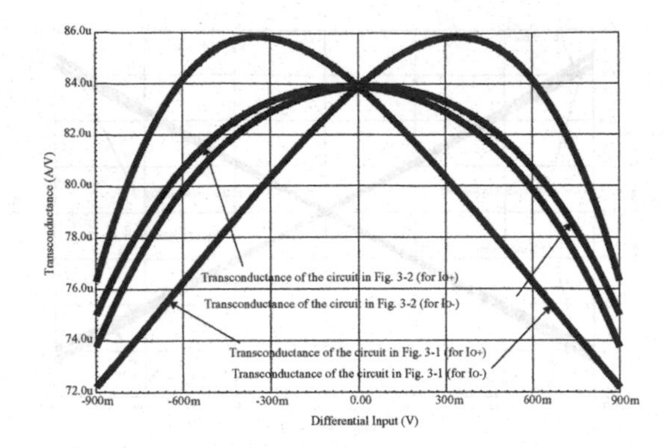

Figure 3-7. G_{m+} and G_{m-} of transconductors in Fig. 3-1 and 3-2 at 400mV control voltage.

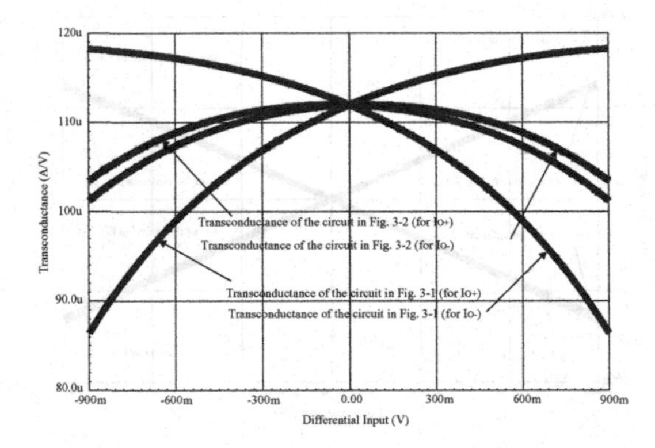

Figure 3-8. G_{m+} and G_{m-} of transconductors in Fig. 3-1 and 3-2 at 1V control voltage.

In order to quantify the linear performance of both transconductors, the following parameter is defined:

$$TE\,(\%) = \frac{G_m(V_{id}) - G_m(0)}{G_m(0)} \times 100 \tag{3-29}$$

where the transconductance error TE defines the linearity of the transconductor's output currents, i.e. the variation in G_m as a function of the input voltage. Figures 3-9, 3-10, and 3-11 show the transconductance error TE for both transconductors at different control voltages. As those figures show, there is a significant improvement in the transconductance variation with input voltage, that is, much lower TE. For instance, if a 3% error is

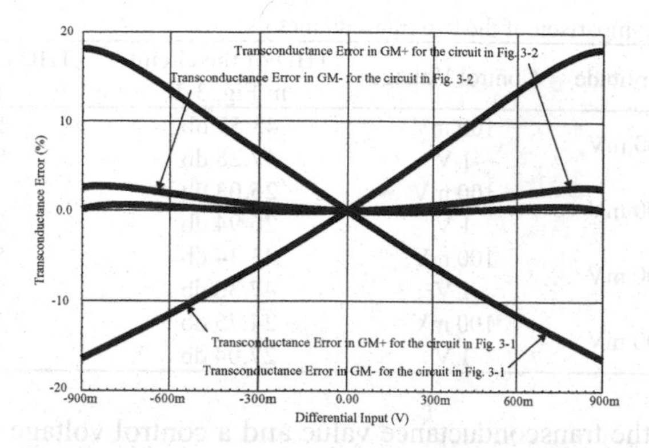

Figure 3-9. TE of transconductors in Fig. 3-1 and 3-2 at 100 mV control voltage.

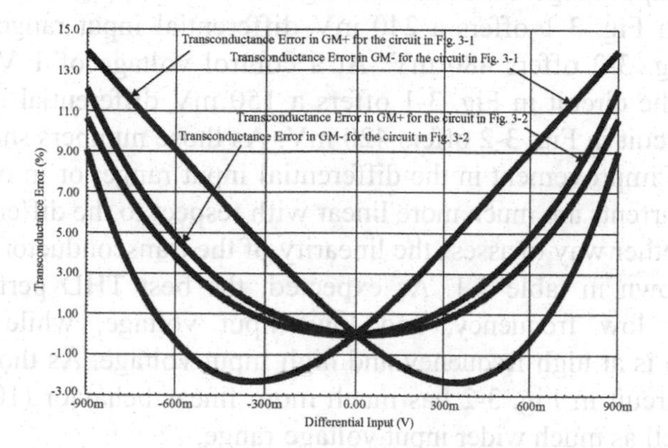

Figure 3-10. TE of transconductors in Fig. 3-1 and 3-2 at 400 mV control voltage.

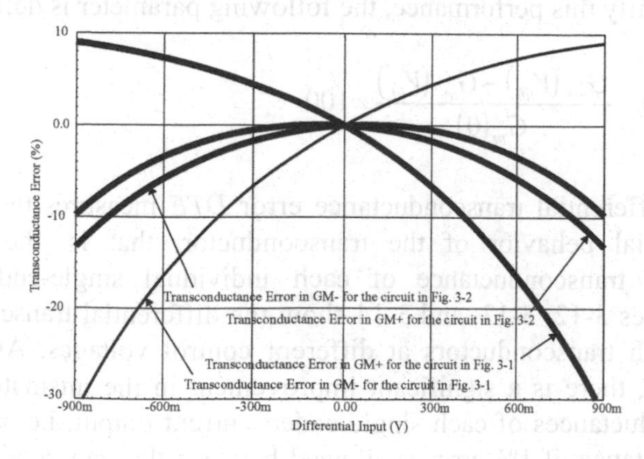

Figure 3-11. TE of transconductors in Fig. 3-1 and 3-2 at 1V control voltage.

Table 3-1. THD comparison of the two transconductors.

Frequency	Amplitude	Control Voltage	THD of the circuit in Fig. 3-1	THD of the circuit in Fig. 3-2
50 KHz	100 mV	100 mV	43.57 db	56.51 db
		1 V	47.28 db	56.38 db
	900 mV	100 mV	25.03 db	50 db
		1 V	26.94 db	39.32 db
50 MHz	100 mV	100 mV	43.34 db	53.18 db
		1 V	47.36 db	56.02 db
	900 mV	100 mV	24.75 db	34.76 db
		1 V	27.04 db	37.79 db

acceptable in the transconductance value and a control voltage of 100 mV is used, Fig. 3-9 shows that the circuit in Fig. 3-1 offers only 120 mV differential input range, while the circuit in Fig. 3-2 offers 900 mV differential input range. At control voltage of 400 mV, Fig. 3-10 shows that the circuit in Fig. 3-1 offers a 240 mV differential input range, while the circuit in Fig. 3-2 offers 480 mV. At a control voltage of 1 V, Fig. 3-11 shows that the circuit in Fig. 3-1 offers a 150 mV differential input range, while the circuit in Fig. 3-2 offers 420 mV. As those numbers show, there is a significant improvement in the differential input range, or in other words, the output currents are much more linear with respect to the differential input voltage. Another way to assess the linearity of the transconductor is the THD numbers shown in table 3-1. As expected, the best THD performance is achieved at low frequency, and low input voltage, while the worst performance is at high frequency and high input voltage. As those numbers show, the circuit in Fig. 3-2 has much more linear behavior (10 dbs better THD), as well as much wider input voltage range.

Another advantage of the circuit in Fig. 3-2 is its truly fully differential nature, or in other words, the minimal difference between G_{m+} and G_{m-}. In order to quantify this performance, the following parameter is defined:

$$DTE\ (\%) = \frac{G_{m+}(V_{id}) - G_{m-}(V_{id})}{G_m(0)} \times 100 \tag{3-30}$$

where the differential transconductance error *DTE* measures the quality of the differential behavior of the transconductor, that is, the difference between the transconductance of each individual single-ended current output. Figures 3-12, 3-13, and 3-14 show the differential transconductance error for both transconductors at different control voltages. As shown in those figures, there is a significant improvement in the mismatch between the transconductances of each single-ended current output, i.e. much lower *DTE*. For instance, if 1% error is allowed between the transconductances of each individual output, then Fig. 3-12 shows that at 100 mV control voltage, the circuit in Fig. 3-1 offers a differential input range of only 30 mV, while the circuit in Fig. 3-2 offers 900 mV of differential input range. At a control voltage of 400 mV, Fig. 3-13 shows that the circuit in Fig. 3-1 offers a 50 mV differential input range, while the circuit in Fig. 3-2 offers 540 mV. At a control voltage of 1 V, Fig. 3-14 shows that the circuit in Fig. 3-1 offers a 30 mV differential input range, while the circuit in Fig. 3-2 offers 330 mV.

Another advantage of the transconductor in Fig. 3-2 is the improvement in control voltage range. Figure 3-15 shows the transconductance versus control voltage for both circuits. As shown, for the circuit in Fig. 3-2, the transconductance value saturates at a higher control voltage, which is predicted by Eqs. 3-27 and 3-28. This implies that the transconductor in Fig. 3-2 has a wider linear control range; a great benefit at low supply voltages.

Noise performance of transconductors is also an important aspect of their design. When digital core transistors with minimum or near minimum channel length are used to achieve higher bandwidth with low power consumption, usually the noise performance is compromised. Since this is exactly the case with the transconductor in Fig. 3-2, the circuit is simulated to asses its noise performance. For this implementation, the integrated input-referred noise from 1MHz to 65 MHz at 1V control voltage is found to be 188 μV, while for the circuit in Fig. 3-1 it is found to be 278 μV. At 100 mV control voltage, the circuit in Fig. 3-2 has 775 μV of integrated input-referred noise, while the circuit in Fig. 3-1 has 1 mV. Note how the circuit in Fig. 3-2 reduces the input-referred noise power by a little less than a half due to its double input stages.

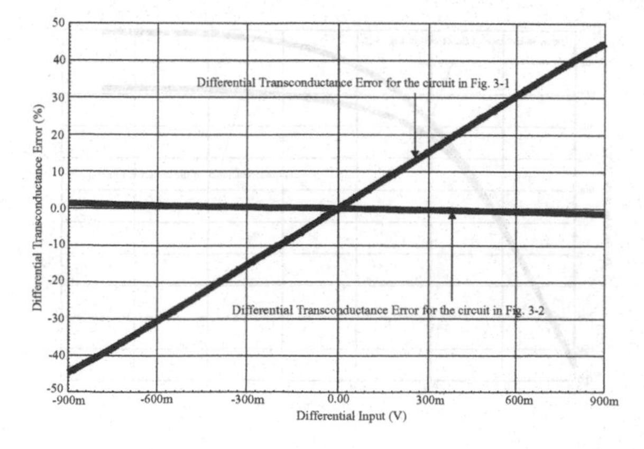

Figure 3-12. DTE of transconductors in Fig. 3-1 and 3-2 at 100 mV control voltage.

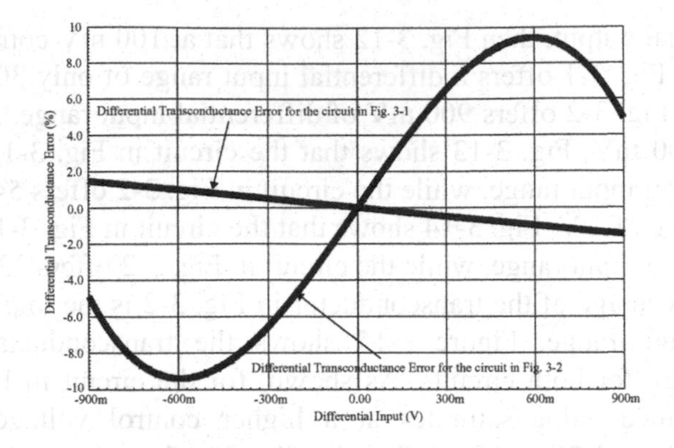

Figure 3-13. DTE of transconductors in Fig. 3-1 and 3-2 at 400 mV control voltage.

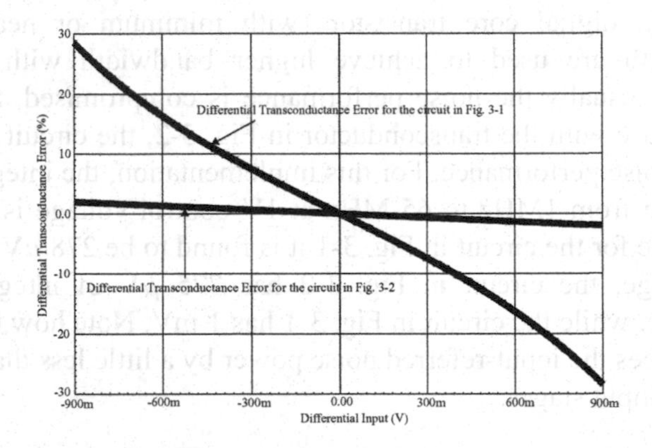

Figure 3-14. DTE of transconductors in Fig. 3-1 and 3-2 at 1V control voltage.

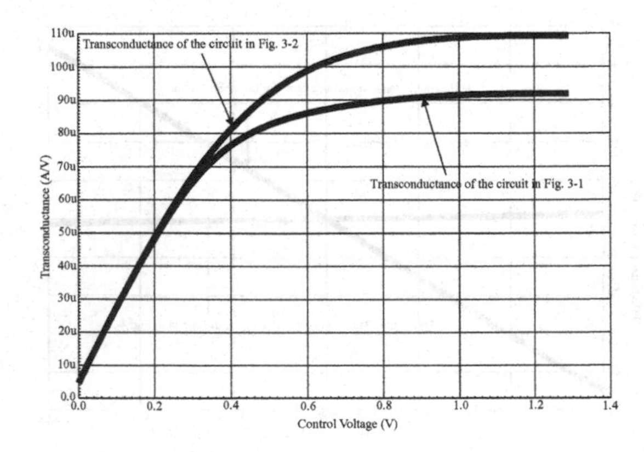

Figure 3-15. G_m versus control voltage of transconductors in Figs. 3-1 and 3-2.

3.2 Comparators

Comparators are very important building blocks in many analog and mixed signal systems, where they mainly serve as an interface between the analog and the digital domain. Thus, comparators are always in the heart of any kind of analog to digital converters design. As the name suggests, a comparator is a circuit block that compares two signals and provides a final digital output, so in essence, a comparator is simply a differential amplifier with a very high gain, where once one of its two input signals slightly exceeds the other, the output of the comparator saturates to one of the two supply rails. There are many parameters that govern the operation of a comparator, and determine its suitability for different applications. Those parameters are mainly gain, resolution, bandwidth, slew rate, and input common-mode range.

The gain of a comparator is defined as the small-signal ratio between the output signal and the differential input signal at DC. Therefore, if the two distinctive digital output levels required are V_H and V_L and the minimum input signal is V_{id}, then the minimum required gain $A_{\min}$ is defined as:

$$A_{\min} = \frac{V_H - V_L}{2V_{id}} \tag{3-31}$$

For example, if the supply rails are $V_H = 3.3\text{V}$ and $V_L = 0$, then for a 5mV differential input signal, the minimum gain required is around 50db. Normally though, the gain is designed to be much higher than the minimum required gain in order to minimize the input-referred offset specially in the case of multi-stage comparators, where the offset introduced by each comparator is divided by the gain of the previous comparator. Hence, the higher the gain is, the lower the total input-referred offset becomes[2].

The resolution of a comparator is the minimum difference between the two input signals that a comparator can resolve and produce a correct digital output for. As opposed to the gain, the resolution is defined with the input-referred offset in place, which means that the resolution of the comparator is not necessarily V_{id} in Eq. 3-31. In fact the resolution is defined to be equal to V_{id} in Eq. 3-31 plus the absolute value of the maximum expected input-referred offset. So, if the maximum expected input offset is 5mV, then using the previous example, the resolution of the comparator is 10mV instead of 5mV as predicted from the minimum gain of the comparator. That obviously gives another justification for increasing the gain of the comparator.

The bandwidth of a comparator is defined as the -3db frequency of the small-signal AC gain response of the comparator. It mainly defines the minimum time the comparator output needs in order to settle at the correct level in response to a small signal input. Hence, the bandwidth determines the maximum speed the comparator could be used for when the input signal is small. The slew rate on the other hand defines the maximum rate the output signal can change by when the input signal is large[2].

Gain, resolution, bandwidth, and slew rate of a comparator are defined using the differential signal at the input with no consideration of the common-mode of that differential signal. Since in practical circuits the common-mode of the input signal is important for defining the operating point of the input circuitry and hence its performance, the input common-mode range is defined as the common-mode range where the comparator can maintain the same gain, resolution, and bandwidth performance. This is particularly important when the input signal has a varying common-mode voltage.

There are numerous methods in the literature for implementing comparators[1,2,5], but those implementations are out of the scope of this book. Instead, a discussion of implementation techniques of a specific class of comparators, i.e. offset comparators, is presented in the next two sections.

3.2.1 Offset Comparators

From the previous definition of a comparator, it compares one of the input signals to the other and produces a digital output accordingly. Another way to look at that definition is that a comparator subtracts the two input signals and compares the result with a predetermined threshold and produces a digital output accordingly. In other words, Eq. 3-31 could be rewritten as:

$$A_{\min} = \frac{V_H - V_L}{2\left(V_{id} - V_{off}\right)} \tag{3-32}$$

where V_{off} is the predetermined offset. In regular comparators, this offset is equal to zero, while in offset comparators this offset is set to a predetermined level. Generally speaking, all comparators are offset comparators in nature since even if the offset is set to zero, the input-referred offset due to input transistors mismatches will cause V_{off} to be non zero. Yet, the expression offset comparator is normally used when the offset is set intentionally by design. One of the most common uses of offset comparators is in the area of signal detection in wire line digital transceivers. It is usually used to differentiate between a valid and an invalid received differential signal on the data lines. For example, in Universal Serial Bus (USB 2.0) transceivers[6,7], a valid signal is a signal that has a differential voltage of at least 125±25 mV. Therefore, an offset comparator is needed to compare the differential signal with a predetermined offset of 125mV to differentiate between a valid and an invalid signal.

Many techniques have been developed in the literature for implementing offset comparators. Figure 3-16a shows a common technique that is often used[8]. In this technique, the circuit adds the predetermined offset by applying a differential offset signal to an extra differential pair identical to the primary input differential pair. The main disadvantage of this technique though is that in order for the offset to be accurate, the differential signal providing the offset has to have the same common-mode as the input signal. Otherwise, due to the finite output impedance of the current sources, the two differential pairs will have different tail currents and consequently different gains, which will lead to significant error in the added offset. This error could vary widely with process, temperature, and supply voltage. In order to maintain the same common-mode for the offset signal and the input signal, a common-mode tracking circuit is needed to extract the common-mode of the input signal and then provide the differential offset signal to the extra differential pair. The problem with common-mode tracking circuits though is that by definition they are required to be slow in order to extract the common-mode of a differential signal without being sensitive to the differential signaling itself. Consequently, a relatively long time will normally be needed for the common-mode tracking circuit to track any change in the common-mode of the signal, during in which the offset loses its accuracy. This limits the usability of this implementation when the input signal has a rapidly varying common-mode.

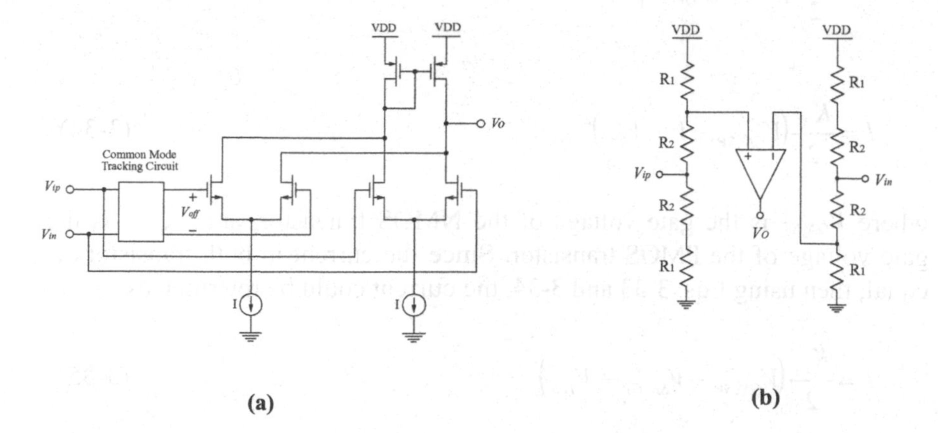

Figure 3-16. Common implementations for offset comparators.

Another technique for implementing offset comparators is shown in Fig. 3-16b. The offset is added by using resistor networks between the supply and ground at the input of a zero-offset comparator. The problem with this implementation though is that the added offset is a function of the supply voltage, which typically could vary as much as ±10%, which compromises the accuracy of the offset. In addition to that, this implementation also compromises the high input impedance nature of the comparator, which is highly undesirable.

In the next section, a technique that uses composite transistors to implement high-speed, accurate, voltage-controlled offset comparators without using a common-mode tracking circuit or compromising the high input impedance nature of the circuit is presented. The comparator is implemented on a standard $0.18\mu m$ digital CMOS process, and simulations performed to verify the design are also presented.

3.2.2 A Voltage-Controlled Offset Comparator

Before presenting the comparator, an introduction to the composite transistor shown in Fig. 3-17 is needed[1,9,10]. The idea behind this block is to use an NMOS transistor on top of a PMOS transistor to form an equivalent transistor with both its gate and source having infinite input impedance. Assuming both transistors are operating in the saturation region, and using the ideal MOS transistor current-voltage equation, the following two equations could be written:

$$I = \frac{K_P}{2}\left(V - V_{SN,GP} - |V_{TP}|\right)^2 \qquad (3\text{-}33)$$

$$I = \frac{K_N}{2}\left(V_{GN,SP} - V - V_{TN}\right)^2 \qquad (3\text{-}34)$$

where $V_{GN,SP}$ is the gate voltage of the NMOS transistor, and $V_{SN,GP}$ is the gate voltage of the PMOS transistor. Since the current in both transistors is equal, then using Eqs. 3-33 and 3-34, the current could be rewritten as:

$$I = \frac{K_{eq}}{2}\left(V_{GN,SP} - V_{SN,GP} - V_{Teq}\right)^2 \qquad (3\text{-}35)$$

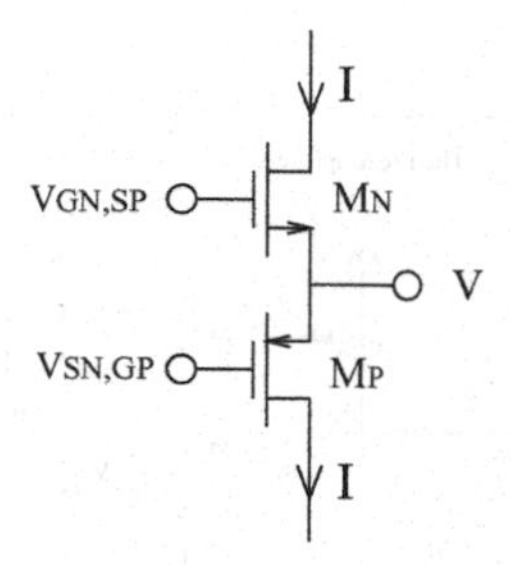

Figure 3-17. The Composite Transistor.

where

$$\frac{1}{\sqrt{K_{eq}}} = \frac{1}{\sqrt{K_N}} + \frac{1}{\sqrt{K_P}} \tag{3.36}$$

$$V_{Teq} = V_{TN} + \left|V_{TP}\right| \tag{3-37}$$

From Eq. 3-35, the composite transistor could be viewed as a single NMOS transistor but with infinite input-impedance at both the gate input $V_{GN,SP}$ and the source input $V_{SN,GP}$, or it could also be viewed as a single PMOS transistor but with infinite input-impedance at both the source input $V_{GN,SP}$ and the gate input $V_{SN,GP}$. The main disadvantage of the composite transistor though is the body effect of the NMOS transistor since its bulk is connected to ground while its source is not (assuming a P-substrate process is used). Due to that, the effective threshold voltage of the composite transistor will vary with the source voltage of the NMOS transistor causing nonlinearity in the behavior of the equivalent transistor, which distorts the output current.

The technique for implementing offset comparators described in this section uses the composite transistor presented earlier to accurately add the required offset without compromising the high input impedance nature of the comparator. Figure 3-18 shows this implementation[11,12]. The circuit is implemented on a standard 0.18μm digital CMOS process with a hybrid 3.3V and 1.8V supplies and both I/O (3.3V) transistors and digital core (1.8V) transistors. The design contains two identical pre-amplification stages followed by a zero-offset comparator (a current mirror OTA and an inverter). The pre-amplification stage adds the required offset to the input differential signal and effectively shifts the threshold voltage from zero to the required offset. It also provides initial amplification to the differential

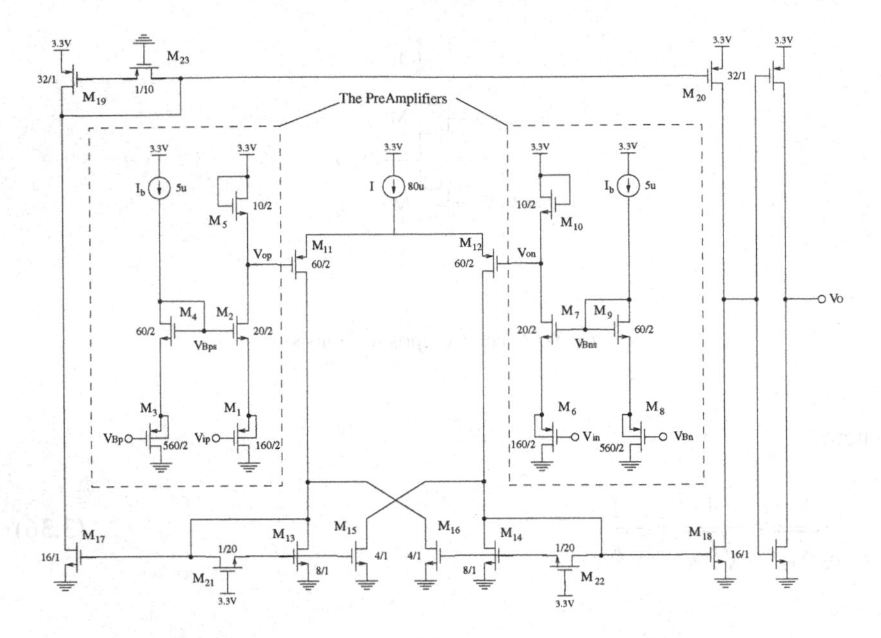

Figure 3-18. A high-speed offset comparator based on composite transistors.

signal in order to enhance resolution and speed. It also acts as a buffering stage between the input data lines and the OTA to avoid kickback effects[2]. Each pre-amplifier is using two composite transistors that act as a single transistor with equivalent V_{Teq} and K_{eq} as shown in Eqs. 3-35 to 3-37. For the pre-amplification stage composed of M_1, M_2, M_3, M_4, and M_5, and assuming that both M_1 and M_2 are operating in saturation-mode, then:

$$V_{Bps} = V_{Bp} + V_{Teq3,4} + \sqrt{\frac{2I_b}{K_{eq3,4}}} \tag{3-38}$$

where $V_{Teq3,4}$ and $K_{eq3,4}$ are defined by Eqs. 3-36 and 3-37 for the composite transistor pair M_3 and M_4, and I_b is a DC biasing current. The current $I_{M1,2}$ flowing in M_1 and M_2 could be written as:

$$I_{M1,2} = \frac{K_{eq1,2}}{2}\left(V_{Bps} - V_{ip} - V_{Teq1,2}\right)^2 \tag{3-39}$$

where again $V_{Teq1,2}$ and $K_{eq1,2}$ are defined by Eqs. 3-36 and 3-37 for the composite transistor pair M_1 and M_2. By substituting Eq. 3-38 into Eq. 3-39, $I_{M1,2}$ could be rewritten as:

$$I_{M1,2} = \frac{K_{eq1,2}}{2}\left(V_{Bp} + \sqrt{\frac{2I_b}{K_{eq3,4}}} - V_{ip} - V_{Teq1,2} + V_{Teq3,4}\right)^2 \quad (3\text{-}40)$$

Since $I_{M1,2}$ is also flowing in M_5, it could be written as:

$$I_{M1,2} = \frac{K_5}{2}\left(3.3 - V_{op} - V_{Tn5}\right)^2 \quad (3\text{-}41)$$

and by equating Eq. 3-40 to 3-41, then V_{op} could be written as:

$$V_{op} = 3.3 - V_{Tn5} - \sqrt{\frac{K_{eq1,2}}{K_5}}\left(V_{Bp} - V_{ip} + \sqrt{\frac{2I_b}{K_{eq3,4}}} - V_{Teq1,2} + V_{Teq3,4}\right) \quad (3\text{-}42)$$

Since the pre-amplifier stage composed of M_6, M_7, M_8, M_9, and M_{10} is identical to the pre-amplifier stage composed of M_1, M_2, M_3, M_4, and M_5, then following the same procedure, V_{on} could be written as:

$$V_{on} = 3.3 - V_{Tn10} - \sqrt{\frac{K_{eq6,7}}{K_{10}}}\left(V_{Bn} - V_{in} + \sqrt{\frac{2I_b}{K_{eq8,9}}} - V_{Teq6,7} + V_{Teq8,9}\right) \quad (3\text{-}43)$$

Since the circuit is symmetric, one can define $K_{eq}=K_{eq1,2}=K_{eq6,7}$, $K_{eq1}=K_{eq3,4}=K_{eq8,9}$, and $K=K_{10}=K_5$, and by subtracting Eq. 3-43 from Eq. 3-42, the following expression for the differential output of the pre-amplifier stage could be written as:

$$V_{op} - V_{on} = E_1 + \sqrt{\frac{K_{eq}}{K}}\left(V_{id} - V_{off} - E_2 - E_3\right) \quad (3\text{-}44)$$

where E_1, E_2 and E_3 are defined as:

$$E_1 = V_{Tn10} - V_{Tn5} \quad (3\text{-}45)$$

$$E_2 = \left(V_{Teq1,2} - V_{Teq3,4}\right) \quad (3\text{-}46)$$

$$E_2 = \left(V_{Teq8,9} - V_{Teq6,7}\right) \quad (3\text{-}47)$$

and $V_{id} = V_{ip} - V_{in}$ is the input differential signal, while $V_{off} = V_{Bp} - V_{Bn}$. Furthermore, if the body effect of transistors M_2, M_4, and M_5 as well as M_7, M_9, and M_{10} is ignored, then E_1, E_2, and E_3 reduce to zero, and Eq. 3-44 could be rewritten as:

$$V_{op} - V_{on} = \sqrt{\frac{K_{eq}}{K_5}}\left(V_{id} - V_{off}\right) \qquad (3\text{-}48)$$

As shown by Eq. 3-48, the differential output of the pre-amplification stage is equal to an amplified, shifted version of the differential input signal V_{id}. This V_{off} shift is set by V_{Bp} and V_{Bn}, which could be adjusted to the required offset accurately (usually through a ratio of resistors, and a band-gap circuit) and without requiring V_{Bp} and V_{Bn} to have the same common-mode voltage as V_{ip} and V_{in}. Note that V_{Bp} and V_{Bn} could be programmable and could be used to compensate for any input-referred offset, thus the name voltage-controlled offset comparator. The amplification factor in Eq. 3-48 could be maximized by making both M_5 and M_{10} smaller than M_2 and M_7. It is also worth mentioning that the effective aspect ratios of the composite transistor pairs M_1-M_2 and M_6-M_7 are a trade off between speed and power. Using high aspect ratios improves speed significantly but increases power consumption since the current flowing into these transistors is not set by a biasing current. In order to maximize the common-mode range of the differential input signal V_{id} that the circuit can handle, V_{Bp} and V_{Bn} need to be as high as possible without pushing M_2 and M_7 into the triode-mode region.

Since in p-substrate based digital CMOS technologies the PMOS transistors are in separate n wells, the input PMOS transistors M_1, M_3, M_6, and M_8 could have their bulks connected to their sources, and hence eliminating their body effect. Yet, NMOS transistors M_2, M_4, M_5, M_7, M_9, and M_{10} have their bulks connected to ground and will suffer from body effect. Therefore, E_1, E_2, and E_3 in Eqs. 3-45, 3-46, and 3-47 become[2]:

$$E_1 = \gamma\left(\sqrt{V_{S10} + |2\phi_F|} - \sqrt{V_{S5} + 2\phi_F}\right) \qquad (3\text{-}49)$$

$$E_2 = \gamma\left(\sqrt{V_{S2} + |2\phi_F|} - \sqrt{V_{S4} + |2\phi_F|}\right) \qquad (3\text{-}50)$$

$$E_3 = \gamma\left(\sqrt{V_{S9} + |2\phi_F|} - \sqrt{V_{S7} + |2\phi_F|}\right) \qquad (3\text{-}51)$$

where γ is the body effect constant and ϕ_F is the Fermi potential. At the switching point of the comparator ($V_{id} = V_{off}$), $V_{S2} - V_{S4}$ is equal to $V_{S9} - V_{S7}$ and $V_{S10} = V_{S5}$, but the errors in Eqs. 3-49, 3-50, and 3-51 are proportional to the square root of the source voltages. Therefore, the errors will not completely vanish, but will tend to be very small. Simulations show that error to be less than 2 mV.

Since the pre-amplifier stage offers only moderate gain, another gain stage is needed to provide the high amplification needed for a comparator. This gain is provided by the current mirror OTA shown in Fig. 3-18 with long-channel active transistor used between the gate and drain of M_{13}, M_{14}, and M_{19} to enhance the bandwidth of the current mirrors and consequently the speed of the OTA[13]. Using transistors M_{21}, M_{22}, and M_{23} instead of passive resistors is justified since the linearity of the resistors is not very critical due to the very large gain of the OTA. The gates of transistors M_{21}, M_{22}, and M_{23} are connected to one of the supply rails to eliminate the need for generating biasing voltages. A positive feedback loop is introduced through transistors M_{15} and M_{16} to enhance the gain of the OTA. Note that the aspect ratios of M_{15} and M_{16} have to be smaller than M_{13} and M_{14} to avoid having hysteresis in the comparator's input-output characteristics[5]. The OTA output stage formed by transistors M_{17}, M_{18}, M_{19}, and M_{20} converts the 3.3V differential output signal of the differential pair to a single-ended 1.8V signal at the final output. Note that M_{19}, M_{20}, and M_{23} are 1.8V transistors, while M_{17}, M_{18}, M_{21}, and M_{22} are 3.3V transistors. This eliminates the need for any special level shifters to interface between analog and digital circuits.

3.2.3 Simulation Results

In order to evaluate the performance of the comparator, AC, DC, and transient simulations are done on the extracted layout of the circuit. The common-mode of the input signal V_{id} is varied between 50mV and 200mV, while the differential amplitude is varied between 50mV and 400mV. The offset V_{off} is set to 125mV by setting V_{Bp} to 400mV and V_{Bn} to 275mV. Simulations show that the whole circuit consumes a total average current of 512 μA and runs at 480 Mbps with power supplies ranging from 3.0V to 3.6V and 1.6V to 2.0V. The temperature range is from -40°C to 125°C. The pre-amplifier stage has unity gain and it consumes 112 □A, while the whole comparator including the OTA has gain of 72.8 db with unity gain frequency of 1.97 GHz. The circuit achieves low power consumption for the operating speed and it maintains the high input impedance nature. It can also track differential signals with rapidly changing common-mode since it doesn't require a common-mode tracking circuit. Transient, DC, and AC simulation results are shown in Figs. 3-19, 3-20, and 3-21 respectively.

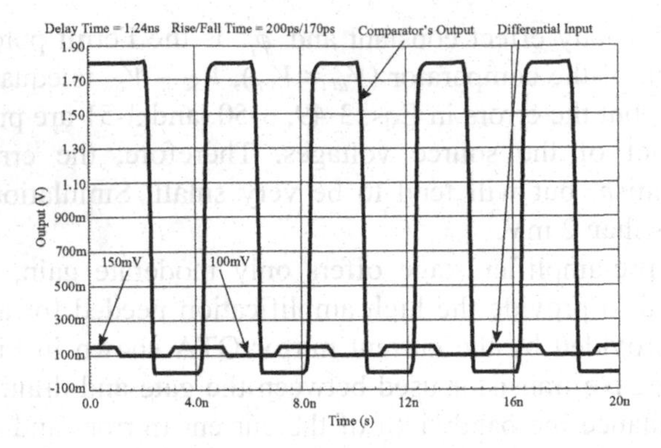

Figure 3-19. Transient response of the comparator in Fig. 3-18.

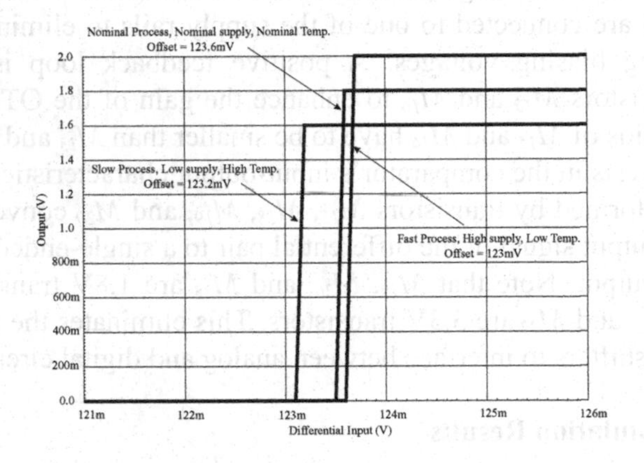

Figure 3-20. DC response of the comparator in Fig. 3-18.

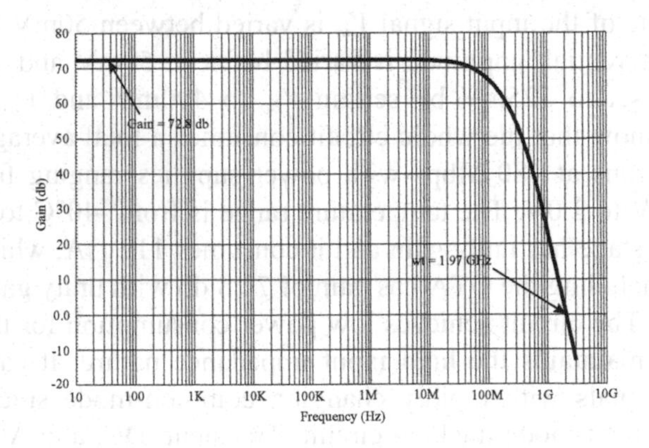

Figure 3-21. AC response of the comparator in Fig. 3-18.

3.3 Summary

This chapter focused on a specific circuit implementation of a low-voltage, high performance, voltage-controlled transconductor. The circuit is a triode-based transconductor that significantly improves the linearity of the output currents with respect to the input voltage, and maintains a wide input/control voltage ranges. Those advantages make this implementation very attractive for low-voltage, highly linear applications. Additionally, the general characteristics of comparators were also discussed with an emphasis on available techniques for implementing offset comparators. A specific circuit design technique for implementing a high-speed voltage-controlled offset comparator was also presented, where higher performance and more accurate control over the offset could be achieved in a very robust fashion.

LIST OF REFERENCES

1. Mohammed Ismail, Terri Fiez, "Analog VLSI Signal and Information Processing," McGraw-Hill, New York, 1994.
2. David A. Johns, Ken Martin, "Analog Integrated Circuit Design," John Wiley & Sons, New York, 1997.
3. Ayman A. Fayed, "Highly-Linear, Wide-Input-Range, Wide Control-Range, Low-Voltage Differential Voltage Controlled Transconductor," US Patent No. 6724258, April 2004.
4. Ayman Fayed, M. Ismail, "A Low-Voltage, Highly-Linear, Voltage-Controlled Transconductor," *IEEE Trans. Circuits Syst. II*, Vol. 52, No. 12, Dec. 2005.
5. P. E. Allen, and D. R. Holberg, "CMOS Analog Circuit Design," Holt, Rinehart and Winston, 1987.
6. Compaq Computer Corporation, Hewlett-Packard Company, Intel Corporation, Lucent Technologies Inc, Microsoft Corporation, NEC Corporation, Koninklijke, Philips Electronics N.V, "Universal Serial Bus Specifications," Revision 2.0, Draft 0.9, Dec. 21, 1999.
7. Don Anderson, "Universal Serial Bus System Architecture," Addison-Wesley, Massachusetts, 1999.
8. D. Yaklin, "Offset Comparator with Common Mode Voltage Stability," US Patent 5517134, May 1997.
9. A.L. Coban, P.E. Allen, "A 1.75 V rail-to-rail CMOS op amp," *IEEE International Symposium on Circuits and systems*, vol. 5, pp. 497-500, May. 1994.
10. C. Galup-Montoro, M.C. Schneider, I.J.B. Loss, "Low output conductance composite MOSFET's for high frequency analog design," *IEEE International Symposium on Circuits and systems*, vol. 5, pp. 783-786, May. 1994.
11. Ayman A. Fayed, and M. Ismail "A High Speed, Low Voltage CMOS Offset Comparator," *Int. J. of Analog Integrated Circuits and Signal Processing,* vol. 36, No. 3, pp. 267-272, Sept. 2003.
12. Ayman A. Fayed, "High speed offset comparator," *US Patent 6400219*, June. 2002.
13. T. Voo, C. Toumazou, "Tunable current mirror technique for high frequency analog design," *IEE Colloquium on Analogue Signal Processing,* pp. 3/1-314, 1994.

Chapter 4

ON-CHIP RESISTORS AND CAPACITORS

Resistors and capacitors are both very important basic building blocks in any analog and mixed signal IC design. Numerous applications in the area of analog signal processing use these two elements, either separately, or combined. They are fundamental elements in oscillators, filters, delay elements, termination networks, amplifiers, and so on. Usually, if simple designs are used to implement any of the above applications, the main characteristics (oscillation frequency of an oscillator, cut-off frequency of a filter ...etc.) of any of them will highly depend on the absolute values of resistors and capacitors used. The accuracy of those elements is always a problem that floats to the surface, especially if the targeted application requires high accuracy for adequate performance. Off-chip discrete resistors and capacitors have high accuracy (typically 1% for resistors and 5% for capacitors), but they have to be used externally. This implies a higher cost and also raises other issues, like the number of pins that can be used to connect those external elements to the internal circuitry. For example, if too many filters are implemented and each one is using a different resistor, it becomes impractical to assign a pin for each one. Also, parasitic elements introduced to the signal path due to bond-wire capacitance and inductance could affect the performance of the circuit causing loss of accuracy intended originally by using external components. The problems introduced by those parasitic elements become more significant at higher speeds.

The ability to integrate resistors and capacitors along with the circuits that use them enables much higher levels of integration and lower cost. CMOS processes offer different ways to implement resistors and capacitors on-chip, but as previously discussed, they suffer from relatively poor control over their absolute values. Most modern CMOS processes can guarantee the accuracy of on-chip resistors and capacitors to only within ±25%. This chapter gives an overview of available techniques to integrate on-chip resistors and capacitors. Different sources of errors in their absolute values due to variations in the fabrication process as well as temperature are presented with an overview of design techniques to minimize those errors. Matching properties of on-chip resistors and capacitors are also discussed along with layout techniques used to improve accuracy and matching.

4.1 Passive Resistors

Passive resistors are implemented using the classic concept of uniform resistivity and geometry of a conducting material. The concept of "passive" here is presented versus the concept of active, which means implementing the resistor function using an active element (a transistor). The universal equation that relates uniform resistivity of a material and its geometry (assuming rectangular geometry) to its total resistance is shown in Eq. 4-1:

$$R = \rho \frac{L}{A} = \rho \frac{L}{W \times t} \quad (4\text{-}1)$$

where ρ ($\Omega.\mu$m) is the resistivity of the material, L (μm) is the length in the direction of current flow, W (μm) is the width perpendicular to the current flow, and t (μm) is the depth of the material[1]. Figure 4-1a shows a diagram of the different dimensions and the layout of an integrated resistor. As Eq. 2-1 shows, the total resistance could be determined by L and W given ρ and t. A modification for Eq. 4-1 is given in Eq. 4-2:

$$R = \rho_S \frac{L}{W} = n \times \rho_S \quad (4\text{-}2)$$

where ρ_S ($\Omega/\square$) is the sheet resistance of the material per square, and n is the number of squares formed by L and W. Taking the resistance of the ohmic contacts into account, the total resistance of the integrated resistor showed in Fig. 4-1b will be:

$$R_{tot} = 2R_{Cont} + n \times \rho_S \quad (4\text{-}3)$$

where R_{Cont} is the contact resistance.

Counting the number of squares is straight forward if the resistor is laid out as a single slap of the conductive material as shown in Fig. 4-1b or as a multiple discontinuous slaps as in Fig. 4-2a. Very often though, when large resistor is needed, it is more efficient area wise to layout the resistor as a serpentine structure as shown in Fig. 4-2b[2]. In this case, current in the corner

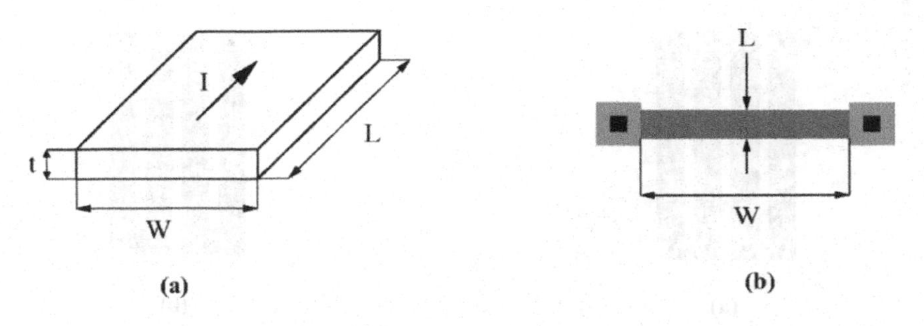

Figure 4-1. (a) Resistor's dimensions, (b) Layout of an integrated resistor.

squares will be changing direction, therefore the simple formula in Eq. 4-2 will not be accurate for estimating the resistance of the corner square. Methods for counting corner squares could be found in the litereture[1], but roughly, corner squares are counted as half squares.

The layers available in CMOS processes for implementing passive resistors are diffusion (n^+, p^+, well) and polysilicon layers[2]. Each type has its own characteristics, behavior with temperature change, and reasons for inaccuracy. As mentioned before, inaccuracy in integrated resistances could be as high as ±25% of the targeted value.

Diffusion is one of the primary methods used to introduce different types of impurities (Boron, Phosphorus ...etc.) to semiconductor materials. Using diffusion, the type of the majority current carriers in silicon, as well as its resistivity can be controlled. The diffusion process starts with introducing a high concentration of the impurity material on the surface of the substrate at high temperature (typically varying from 900 to 1200 °C depending on the type of impurity) for a specific period of time. The impurity material will then diffuse to a certain depth inside the substrate causing its electrical characteristics to change. Figure 4-3 shows the diffusion process of a donor material into a P-substrate. The donor material diffused into the P-substrate converts the p-type silicon to n-type silicon, and therefore changes the majority current carries, which essentially changes the resistivity of silicon. The point where the type of silicon changes from n-type to p-type is called the junction depth, which is a function of the diffusion process, i.e. concentration of impurities introduced, as well as diffusion time and temperature[1].

As shown in Eq. 4-2 the total resistance of the diffused material could be calculated given the sheet resistance and the number of squares. The sheet resistance could be calculated if the resistivity of the conducting material is

Figure 4-2. (a) Multiple fingers resistor, (b) Layout of a serpentine structure.

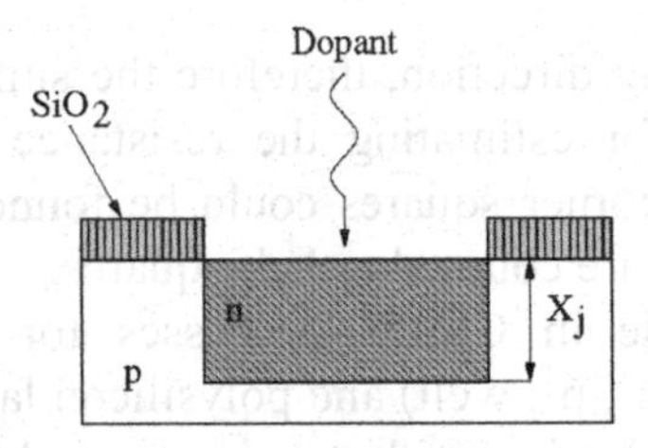

Figure 4-3. The diffusion process.

uniform across the depth of the material, however, this is not the case in the diffusion process. In that case, the sheet resistance can be written as[1]:

$$\rho_S = \frac{1}{\int_0^{X_J} q\mu N(X)dx} \tag{4-4}$$

where X_J is the junction depth, μ is the majority carrier mobility, and $N(X)$ is the net impurity concentration as a function of depth. It can be shown that for a given diffusion profile, the sheet resistance ρ_S is uniquely related to the surface concentration of the diffused material as well as the background concentration of the substrate[1]. Depending on the type and the concentration of the diffused impurity, the sheet resistance of the diffused resistor could be in the order of KΩ/□ like in the case of well resistors (n or p type), or it could be in the order of tens of Ω/□ like in the case of source and drain diffusion resistors (n^+ or p^+).

The polysilicon layer is usually used to form the gates of transistors. It is made up of small crystalline regions of silicon. Therefore, in the strictest sense, polysilicon is not amorphous silicon, and it is not crystalline silicon such as the wafer[3]. Amorphous silicon is made up of randomly organized silicon atoms, while crystalline silicon is formed by silicon atoms organized in an orderly fashion. The use of polysilicon gates is a key advance in modern CMOS technologies since it allows the source and drain regions to be self-aligned to the gate, thus eliminating parasitics from overlay errors between the gate and the source and drain diffusion. The polysilicon layer used for the gates however has to be highly doped, otherwise the depletion region of the gate itself will result in a series capacitance with the gate oxide capacitance, which in turn leads to a reduced inversion layer charge density and degradation of the MOSFET transconductance[4]. Therefore, polysilicon has a very low resistivity in the order of 20 to 30 Ω/□, and it can be used to implement relatively low resistors.

A special type of polysilicon resistors is known as silicide-block poly resistors. This type of polysilicon resistors requires an extra mask during the fabrication process to block the high-doping step performed on polysilicon used for the gates of transistors from affecting the silicide-block polysilicon. Therefore, the doping of silicide-block resistors is kept around a critical value that minimizes the variations in the sheet resistance of that layer with temperature. This issue will be discussed later in detail. The critical sheet resistance of silicide-block polysilicon at which a minimum variation with temperature could be achieved is around 300 Ω/□.

Diffusion resistors, specifically well-diffusion, have the advantage of having a higher sheet resistance, therefore implementing the required resistance will consume a relatively smaller area than polysilicon resistors. It suffers though from poor noise and speed performance due to its direct contact with the substrate. This direct contact with the substrate injects noise from other circuits on the same substrate into the body of the resistor. The parasitic capacitance between the resistor and the substrate is also significant. One way to reduce the noise coupling from the substrate is the use of a grounded guard-ring around the well-diffusion to absorb the noise from the substrate[2]. Figure 4-4 shows the guard-ring. It is also worth mentioning that well resistors have a slight voltage-dependent behavior. The reason behind this behavior is that the depletion region depth in the well resistance is a function of the reverse bias between the substrate and the common-mode voltage of the plus and minus terminals of the resistor. Since the extension of the depletion region into the well reduces the effective depth of the well, it causes the sheet resistance to slightly change with voltage, and therefore causing some nonlinearity in the resistor performance.

Polysilicon resistors have the advantage of being isolated from the substrate, which enhances their noise performance. Additionally, this isolation from the substrate implies lower parasitic capacitance, and hence

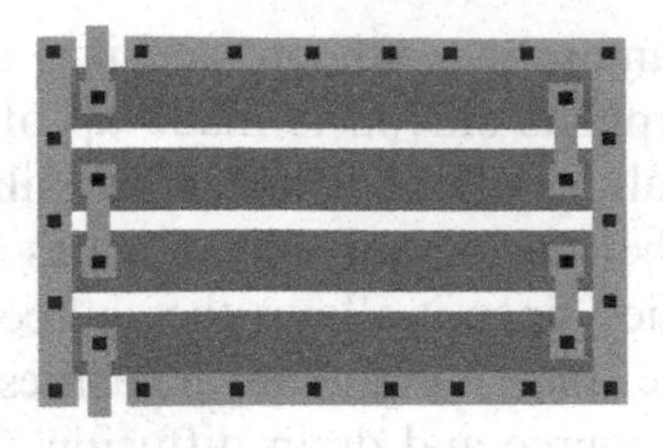

Figure 4-4. A well-resistor with a guard-ring to reduce noise coupling from the substrate.

better speed performance. Furthermore, polysilicon resistors do not exhibit a voltage-dependent behavior as opposed to diffusion resistors. Silicide-block polysilicon resistors specifically, have the advantage of minimal variation with temperature, yet they require extra processing steps during fabrication.

Generally, polysilicon resistors have the disadvantage of having low sheet-resistance, which implies higher area consumption to implement the required resistance.

4.1.1 Accuracy of Passive Resistors

In the development of any process, it is very important to evaluate how well the process can control performance parameters of different devices. For that purpose, test structures of devices of interest are built and then measured for different wafers and lots. This evaluation process results in a statistical distribution for the parameters of interest across different process conditions. The test structures used have to be simple to minimize measurement errors, yet they have to be a good and accurate representative of the different parameters under test. The task of building those test structures might be complicated especially if too many parameters need to be characterized. In the case of resistors, typically van der pauw structures are used to measure the sheet resistance ρ_S, a pair of resistors with similar lengths to determine the encroachment, and a Kelvin contact resistance structure to determine the combined contact and spreading resistance introduced by a contact. There have been also other proposed structures in the literature that can closely determine the accuracy and matching properties of on-chip resistors across process variations[3].

As Eq. 4-2 suggests, the error in the value of R is a function of the error in ρ_S, L, and W. The First source of error is ρ_S, which is mainly a function of the fabrication process, and temperature variations during the circuit operation. During fabrication, the uncertainty in the diffusion time, temperature, and doping concentration determines the amount of error in the absolute value of ρ_S. This error varies from one process to another depending on how well the different processing steps are controlled. Usually, good characterization of different resistor structures on a given process node yields an accurate estimation of the amount of error expected. On the other hand, after the fabrication process is over, and a value of ρ_S is maintained to within certain accuracy, during circuit operation the temperature of the resistor's material changes as a function of the current flow in the resistor. It is a well known fact that the resistivity of any conducting material varies with temperature, and therefore additional error is added to the absolute value of the resistor due to variations in ρ_S with temperature[1]. The variation with temperature is more difficult to account for since it changes with time and circuit operation, unlike errors introduced by the fabrication process, where once they happen they stay the same. Generally speaking, the circuit designer has to make sure that his circuit can tolerate those variations, otherwise a tuning scheme has to be continuously monitoring the value of the resistor and correcting for temperature related errors.

A typical measure of variations in the resistor value with temperature is defined by the temperature coefficient[4]:

$$TC = \frac{1}{R}\frac{\partial R}{\partial T} \tag{4-5}$$

where R is the resistance value at nominal temperature, and T is the temperature in Kelvin. TC is usually measured in ppm (parts per million).

Generally, the sign of the temperature coefficient (+ or -) is a function of the current carriers concentration in room temperature. When temperature increases, more energy becomes available to electrons (or holes) to be liberated from the nucleus and contribute to the current flow, hence reducing the effective total resistivity of the resistor, i.e. negative temperature coefficient. On the other hand though, increasing temperature leads to more vibrations in the atoms of the conductor's material, which consequently reduces the mean-free-path current carriers can travel before they hit the lattice and lose their energy, i.e. phonon-lattice scattering. This reduction in the mean-free-path of current carriers increases the resistivity of the resistor, i.e. positive temperature coefficient. Since there are two contradicting factors, the effective temperature coefficient will depend on which factor is more dominant. Generally, if current carriers' concentration at room temperature is already high, the second factor tends to dominate leading to a positive temperature coefficient, while if this concentration is low, the first factor tends to dominate leading to a negative temperature coefficient. Since doping level determines the concentration of current carriers, then by changing the doping level, the temperature coefficient could be positive, negative, or possibly zero.

Diffusion resistors in general tend to have a positive temperature coefficient. Well-resistors have a temperature coefficient ranging from 1500 to 2500 ppm. Polysilicon resistors on the other hand have two different types. The first type is the one used for the gates of MOS transistors. This type is highly doped and has very low resistivity in the order of 20 to 30 Ω/□ . Since it is highly doped (higher concentration of carriers), this type has a positive temperature coefficient ranging from 1000 to 2000 ppm. The second type is the silicide-block polysilicon resistor, where the doping of silicide-block resistors is kept around a critical value that yields almost zero, or very low temperature coefficient. The critical sheet resistance of silicide-block polysilicon at which very low temperature coefficient could be achieved is around 300 Ω/□. This type of resistors usually has a temperature coefficient ranging from -10 to 10 ppm.

The second source of error in the resistor value is the error in *W*, and *L*. This is mainly caused by the etching process during the photo-step that defines the geometry of the resistor. Figure 4-5 shows the etching process for a well-resistor and a ploy-resistor. As shown in Fig. 4-5a, the actual width (or length) of the well area, where diffusion takes place, is wider than the drawn width on the photo mask, therefore the effective resistance of the well area could be higher or lower than expected depending on the error ΔW and ΔL. The same concept also applies to polysilicon resistors as shown in Fig. 4-5b with the difference that actual *W* or *L* is always less than the drawn ones. A unique phenomenon that happens only in diffusion resistors is the lateral diffusion process. As shown in Fig. 4-5a, the diffused material tends to diffuse laterally. Therefore the effective width (or length) of the well-resistor will be higher than the original drawn one. This phenomenon adds to ΔW and ΔL as well. Taking into account the errors in ρ_S, *W*, and *L*, the total resistance could be rewritten as:

$$R = \left(1 + \frac{\Delta\rho_S}{\rho_S}\right) \frac{\left(1 + \frac{\Delta L}{L}\right)}{\left(1 + \frac{\Delta W}{W}\right)} \rho_S \frac{L}{W} = EF * R_{\text{Ideal}} \tag{4-6}$$

where $R_{\text{Ideal}} = \rho_S\ (L/W)$ is the required resistance value, and *EF* is the error factor due to errors in ρ_S, *W*, and *L*. The percentage error in the resistor value could then be written as:

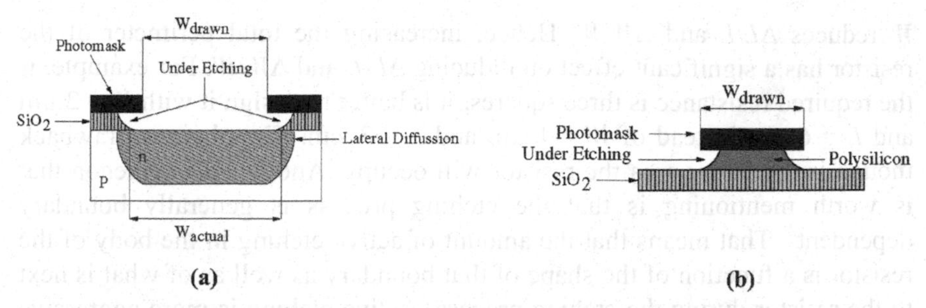

Figure 4-5. (a) Etching and lateral diffusion errors in well resistors, (b) Etching errors in polysilicon resistors.

$$Error(\%) = [EF - 1] \times 100 \tag{4-7}$$

As shown by Eq. 4-7, in order to minimize the error in the resistance, the error factor EF has to be as close as possible to 1. Analysis of Eq. 4-6 suggests that in order to minimize the error in the resistor, $\Delta\rho_S/\rho_S$, $\Delta W/W$, and $\Delta L/L$ need to be minimized.

In order to minimize $\Delta\rho_S/\rho_S$, $\Delta\rho_S$ needs to be minimized, and ρ_S needs to be maximized. Since ρ_S is set by the process, nothing could be done from design or layout perspective to maximize it. This leaves us with minimizing $\Delta\rho_S$, which generally has random and systematic components. The random component of $\Delta\rho_S$ results from random spatial variations in process parameters, while the systematic component results from systematic gradients in process parameters. The random component could be greatly reduced if the total area of the resistor is increased simply because the random component tends to have a Gaussian distribution with a zero average. Hence, the larger the area of the resistor is, the closer the effective value of the random component of $\Delta\rho_S$ is to zero and consequently a lower $\Delta\rho_S$. The systematic component of $\Delta\rho_S$ on the other hand tends to increase with increasing the area since the larger the area span of the resistor is, the larger the effect of process gradients on the systematic component of $\Delta\rho_S$, and consequently a higher $\Delta\rho_S$. Therefore, there is a ceiling to the improvement in $\Delta\rho_S$ that could be achieved by increasing the area, and beyond that ceiling $\Delta\rho_S$ starts to degrade.

On the other hand, $\Delta L/L$ and $\Delta W/W$ could be minimized by simultaneously reducing ΔL and ΔW and increasing L and W respectively. Since ΔL and ΔW are functions of the etching process, again they have systematic and random components. Therefore, increasing L and W tends to reduce the random components of ΔL and ΔW, which also adds to the fact that increasing L and

W reduces $\Delta L/L$ and $\Delta W/W$. Hence, increasing the total perimeter of the resistor has a significant effect on reducing $\Delta L/L$ and $\Delta W/W$. For example, if the required resistance is three squares, it is better to design it with $W = 2\ \mu$m and $L = 6\ \mu$m instead of $W = 1\ \mu$m and $L = 3\ \mu$m. The obvious drawback though is the larger area the resistor will occupy. Another phenomenon that is worth mentioning is that the etching process is generally boundary dependent[2]. That means that the amount of active etching in the body of the resistor is a function of the shape of that boundary as well as of what is next to the resistor during the etching process. Active etching is more aggressive if there is empty space around the boundary of the resistor. Figure 4-6a shows an example of how ΔW (or ΔL) could be different to the left of the resistor as opposed to its right due to different boundary condition. In addition to the fact that ΔW (or ΔL) is higher if there is empty space on the boundary, the non uniformity in ΔW (or ΔL) across the structure of the resistor causes additional uncertainty in its value. Thus, it is a good practice in precise applications to add dummy resistors around the main resistor as shown in Fig. 4-6b to guarantee less active etching and uniform boundary conditions, hence a lower and uniform ΔW and ΔL. Another problem that is related to the layout of the resistor in Fig. 4-2b is that since the etching process is boundary dependent, there is higher active etching at the edges of corner squares[2]. This causes the edges of corner squares to be rounded leading to poor control over their shape. The poor control over the shape of corner squares causes uncertainty in their contribution to the total resistance. For precise applications, sharp edges should be designed to be rounded or with 45 degrees as shown in Fig. 4-7. This way, good control over the shape of corner squares could be maintained leading to more accurate resistors.

Figure 4-6. (a) Boundary dependent Etching, (b) Etching with dummies present.

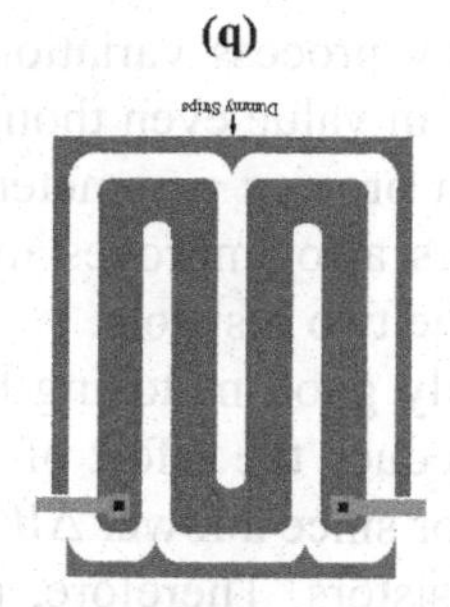

Figure 4-7. (a) Resistor layout with dummy strips and (a) 45° corners, (b) rounded corners.

4.1.2 Matching Properties of Passive Resistors

As discussed in the previous section, uncertainties in the absolute value of integrated resistors could be as high as ±25%. This loose accuracy might be inadequate for many applications. Matching between resistors on the other hand is much more accurate than the absolute value of each individual resistor. In fact, with careful layout, matching between resistors could be maintained with less than 0.1% error for some processes[2]. Generally speaking, the reason for the ability of the fabrication process to achieve good matching is that variations in process parameters from one point to another on the die is likely to be very small if these two points are fairly close to each other. Therefore, if two resistors are in very close proximity, it is very likely that any change in process parameters within the area these two resistors occupy is minimal. It is worth mentioning that the ability of the fabrication process to match two elements applies not only to resistors, but also for transistors and capacitors. This fact has driven the development of many circuit techniques that depend solely on matching rather than absolute values; switch capacitor circuits are good examples.

There are many design and layout guidelines for maintaining good matching between integrated resistors. Generally, if two resistors are close to each other, then process parameters for both of them will be very close, yet, this is not enough to achieve the best matching. The layout has to also guarantee that the two resistors are affected the same way by any change in process parameters even if this change is similar for both. There are different levels of layout complexity in order to achieve good matching between two resistors. The first level is simply laying out the two resistors in a very close proximity and keeping the same orientation for both. Figure 4-8 shows a good and a poor layout. Keeping the same orientation is likely to cause the two resistors to be affected the same way by any process variations. This means that the two resistors will track each other in value even though their absolute value is very sensitive to any change in process parameters. Note that adding dummy elements around the resistors also improves matching since it guarantees a uniform etching rate across the two resistors.

Generally, even if the application requires only good matching between two resistors, it is still a good practice to try to reduce the effect of ΔW and ΔL on the absolute value of each individual resistor since a lower ΔW and ΔL means also lower mismatch between the two resistors. Therefore, it is not recommended to use minimum dimensions allowed by the process for either L or W, and using larger dimensions is preferable, if area allows. However, one has to be careful because even though increasing L and W reduces mismatch due to lower ΔW and ΔL, area increase associated with that might increase systematic errors in ρ_S due to process gradients over large areas.

Sometimes, it is not possible to design each resistor as a single slap of material as shown in Fig. 4-8, specially if the resistor value is too high,

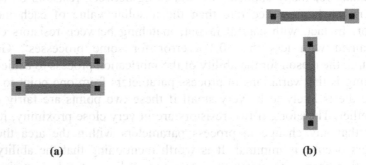

Figure 4-8. Two matched resistors: (a) good layout, (b) poor layout.

Figure 4-9. Layout of two matched interdigitized resistors.

which requires very long structures, or too low, which requires very wide structures. In that case, laying out the resistor as a single slap of material will suffer from long range process variations and therefore poor matching. In these cases laying out the resistors in a serpentine structure as shown in Fig. 4-2b is very common. The serpentine layout suffers from smaller $\Delta\rho_S$ in both X and Y dimensions, which usually yields lower overall error in the value of the resistor than large $\Delta\rho_S$ in only one dimension. On the other hand, matching between two serpentine resistors is not that good if the two resistors are simply laid out next to each other. That leads to the second level of complexity in the layout of two matched resistors, which is the interdigitized approach. In this approach, each resistor is designed using multiple fingers instead of a serpentine structure, then the two resistors are laid out as shown in Fig. 4-9. In the interdigitized structure, the centroids of both resistors are kept very close to each other, and therefore better matching is achieved, usually within 0.5%[2]. As shown in Fig. 4-9, at each given point on the X dimension there is one finger from one resistor and another finger from the other resistor at very close proximity. Hence, both resistors will suffer equally from process variations in the X dimension. However, the interdigitized approach does not account for variations in the Y dimension. Usually this is not a problem if the size of each resistor finger in the Y direction is short enough to neglect process variations in the Y dimension. If this is not the case, then the only solution is to divide each resistor into more fingers to reduce the span in the Y direction.

The third level of complexity in the layout of the two matched resistors is the common-centroid (also called cross-coupled) approach shown in Fig. 4-10. In this approach, each resistor is divided into two elements and then laid out as shown in Fig. 4-10a. Starting from the center of the common-centroid structure, variations in process parameters in both the X and Y directions are always seen by an element of each resistor. Thus, the two resistors will suffer equally from process variations in both dimensions. If the number of fingers that compose each resistor element in Fig. 4-10a is designed to be an even number, then yet another level of accurate matching could be achieved by interdigitizing each resistor element as shown in Fig. 4-10b.

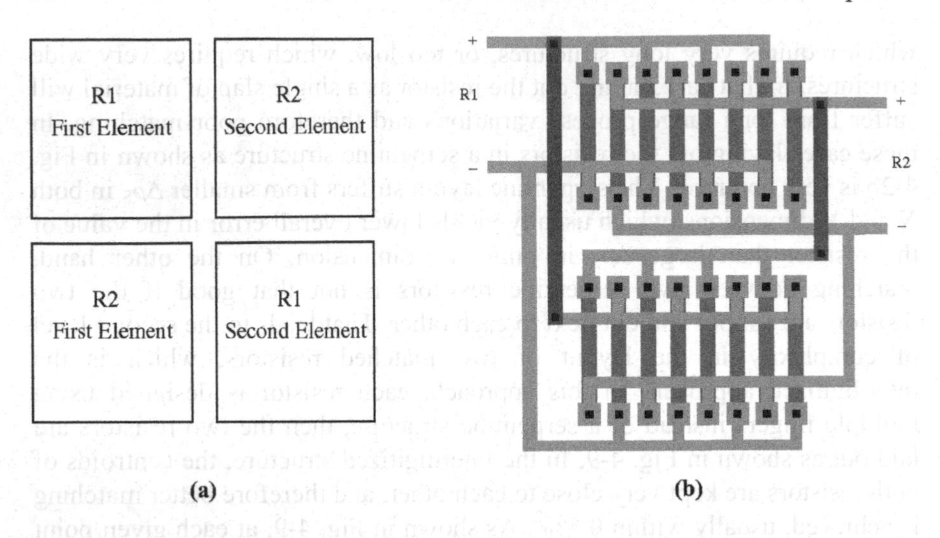

Figure 4-10. (a) Cross-coupled layout, (b) Layout of two matched common-centroid interdigitized resistors.

4.2 Passive Capacitors

Parallel plate capacitors are the most common form of capacitors, where two plates of a highly conductive material with a dielectric in between are simply all what is needed to implement the capacitor. The value of the capacitor can be determined using the following equation:

$$C = \frac{\varepsilon_o \varepsilon_r}{t} WL = C_S A \tag{4-8}$$

where $\varepsilon_o \varepsilon_r$ (fF/μm) is the permittivity of the dielectric material (3.45×10^{-6} fF/μm for silicon dioxide), t (μm) is the dielectric thickness, and WL (μm^2) is the area of the capacitor plate[2]. The specific capacitance C_S (fF/μm^2) is the capacitance per unit area or per square. CMOS technologies naturally enable the implementation of capacitors. Two of the highly conductive layers available in the process could be used to implement the top and bottom plates of the capacitor with the silicon dioxide layer in between acting as a dielectric. Those conductive materials are the diffusion layers (p^+, n^+), the polysilicon layer, and the metal layers. Before discussing different capacitors available in CMOS processes, the fundamentals of Metal-Oxide-Semiconductor (MOS) capacitors are discussed since computing the capacitor value is more complicated than the formula presented in Eq. 4-8.

A simple MOS structure is shown in Fg. 4-11. Here, the semiconductor or the well is chosen to be P-type silicon, but the analysis is the same for N-type silicon. For simplicity the well is assumed to be at zero potential, while the metal potential is V_g. When V_g is negative, more holes are attracted to the well-oxide interface, which effectively increases the conductivity of the well right beneath the oxide. This condition is called the "accumulation" condition since the majority current-carriers (holes in this case) are being accumulated at the well-oxide interface due to the negative voltage applied to the metal. Since the conductivity of the well right beneath the oxide is significantly high during accumulation, any charge stored on the capacitor is practically stored in a very thin layer right beneath the oxide. This means that the effective capacitance of the structure is essentially the oxide capacitance C_{ox} and can be determined using Eq. 4-8. When V_g is equal to zero, no accumulation of majority carriers happens right beneath the oxide, i.e. low conductivity. Therefore the charge stored on the capacitor is distributed across a certain depth in the well, which effectively adds a series capacitance to C_{ox}. This condition is referred to as the "flat-band" condition. The flat-band capacitance is essentially smaller than C_{ox} and can be determined by[5]:

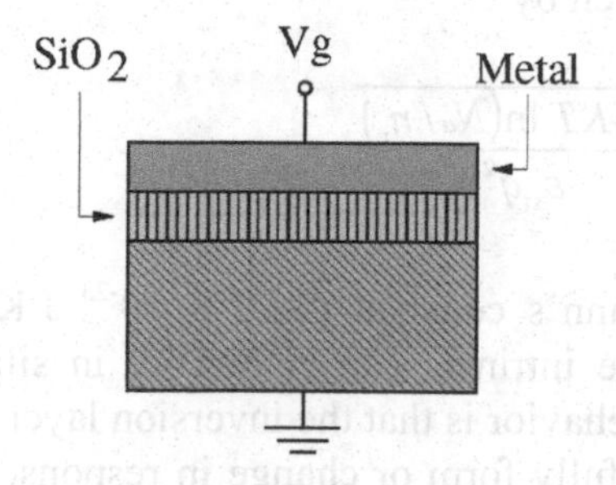

Figure 4-11. A MOS Structure.

$$\frac{1}{C} = \frac{1}{C_{ox}} + \sqrt{\frac{KT}{\varepsilon_{si} q^2 N_a}} \tag{4-9}$$

where ε_{si} is the silicon permittivity (1.04×10^{-12} F/cm), q is the charge of the electron (1.6×10^{-19} C), and N_a is the doping concentration. When V_g starts to slightly turn positive, holes start to be repelled away from the well-oxide interface due to the positive voltage applied to the metal, which creates a depletion region beneath the oxide, and thus, this condition is referred to as the "depletion" condition. In the depletion condition the charge stored on the capacitor is distributed across the depth of the depletion region created. This distributed storage of charge can be modeled as an extra capacitor in series with the oxide capacitor C_{ox}, which causes the effective capacitance of the MOS structure to decrease. The total value of the capacitance in the depletion condition could be shown to be[5]:

$$C = \frac{C_{\mathrm{ox}}}{\sqrt{1 + \left(2C_{\mathrm{ox}}^2 V_g / \varepsilon_{\mathrm{si}} q N_a\right)}} \tag{4-10}$$

As V_g increases, an inversion layer of minority carriers (electrons in this case) starts to form at the well-oxide interface, which effectively again increases the conductivity of the well layer right beneath the oxide. This condition is referred to as the "inversion" condition. In this condition, any charge stored on the capacitor is practically stored in the thin layer right beneath the oxide, which brings the total capacitance back to C_{ox}, exactly as it was during the accumulation condition. It is worth mentioning though that if the variation in V_g is fast (100 Hz or more), the total capacitance during the inversion condition does not rebound to C_{ox}. In fact, it keeps decreasing until it hits a C_{min} given by[5]:

$$\frac{1}{C_{\mathrm{min}}} = \frac{1}{C_{\mathrm{ox}}} + \sqrt{\frac{4KT \ln\left(N_a / n_{\mathrm{i}}\right)}{\varepsilon_{\mathrm{si}} q^2 N_a}} \tag{4-11}$$

where K is Boltzmann's constant (1.38×10^{-23} J/K), T is temperature in Kelvin, and n_i is the intrinsic carrier density in silicon. The fundamental reason behind that behavior is that the inversion layer right beneath the oxide needs some time to fully form or change in response to variations in V_g. If the variations in V_g are relatively fast, then the inversion layer will not be able to fully respond. This phenomenon increases the effective depth of the inversion layer, and therefore forms a series capacitance with C_{ox}. This series capacitance keeps the total value of the capacitance at the lower value shown in Eq. 4-11. Figure 4-12 shows the capacitance variation of a MOS structure with a P-type well versus V_g for both low and high frequencies. N-type MOS structures exhibit the same capacitance-voltage behavior with an inverted metal-to-well voltage polarity. Analyzing Fig. 4-12, one can conclude that in order to get the maximum capacitance from a MOS capacitor, it has to be biased into either the accumulation or the inversion region, while the depletion condition has to be avoided. However, if the metal voltage V_g is a

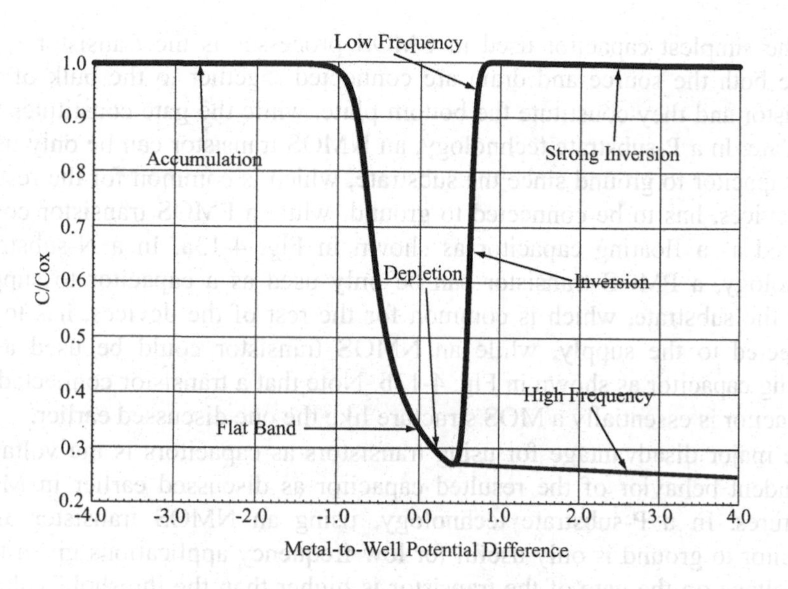

Figure 4-12. C-V characteristics for a MOS Structure.

high frequency signal, then the only frequency-reliable region is the accumulation region. The only limitation of the accumulation region though is that in order to get the full C_{ox}, the capacitor has to be biased at around 1V more negative than the flat-band condition since the flat-band capacitance is a little less than C_{ox} as shown in Eq. 4-9. A workaround this problem could be found by analyzing Eqs. 4-9 and 4-10. The more the doping (N_a) of the well is, the closer the flat band capacitance and the depletion capacitance are to C_{ox}. Therefore, by increasing N_a, the dip in the C-V curve shown in Fig. 4-12 between the accumulation and the inversion regions will decrease, which effectively reduces the restriction on how negative (or positive) the metal-to-well voltage has to be in order to get the total capacitance to C_{ox} in both the accumulation and inversion regions. This is especially very important to maximize the constant-capacitance voltage range. It is worth mentioning that even though biasing the MOS capacitor in the depletion region is generally not desirable, but has a very interesting application as a voltage controlled capacitor (a varactor). So it could be used to tune the frequency of an oscillator or the center frequency of a filter, and so forth. Having reviewed the fundamentals of MOS capacitors, a discussion of different options for implementing capacitors in CMOS technologies will now be presented.

The simplest capacitor used in CMOS processes is the transistor itself where both the source and drain are connected together to the bulk of the transistor and they constitute the bottom plate, while the gate constitutes the top plate. In a P-substrate technology, an NMOS transistor can be only used as a capacitor to ground since the substrate, which is common for the rest of the devices, has to be connected to ground, while a PMOS transistor could be used as a floating capacitor as shown in Fig. 4-13a. In a N-substrate technology, a PMOS transistor can be only used as a capacitor to supply since the substrate, which is common for the rest of the devices, has to be connected to the supply, while an NMOS transistor could be used as a floating capacitor as shown in Fig. 4-13b. Note that a transistor connected as a capacitor is essentially a MOS structure like the one discussed earlier.

The major disadvantage for using transistors as capacitors is the voltage-dependent behavior of the resulted capacitor as discussed earlier in MOS structures. In a P-substrate technology, using an NMOS transistor as a capacitor to ground is only useful for low frequency applications given that the voltage on the gate of the transistor is higher than the threshold voltage of to ensure biasing in the inversion region, hence yielding the highest constant capacitance possible. Therefore, if the gate-to-bulk potential difference across the transistor is less than the threshold voltage, then using NMOS transistors as capacitors is not desirable due to voltage-dependent capacitance in depletion region. Note that NMOS transistors can not be biased in accumulation since their bulk has to always be connected to the lowest voltage in the circuit. A PMOS transistor on the other hand could be used as a floating capacitor. If the gate-to-bulk voltage is positive, the capacitor will be biased at the accumulation region, hence the capacitance will be equal to C_{ox}. If the gate-to-bulk voltage is negative (more than the

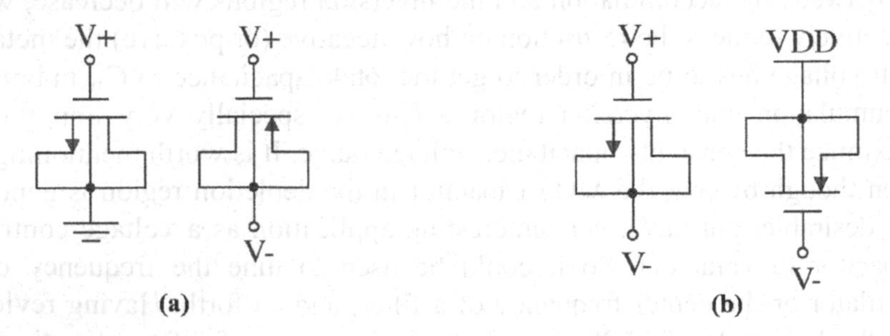

Figure 4-13. Transistors used as capacitors: (a) P-substrate, and (b) N-substrate technologies.

threshold voltage), the capacitor will be biased at the inversion region and the capacitance will be equal to C_{ox} only at low frequencies. As mentioned before, the reliable region for all frequencies in MOS structures is the accumulation region, therefore if the application using the capacitor can guarantee that the potential difference across the capacitor will always be positive, then using a PMOS transistor with its gate as the plus side of the capacitor and its bulk as the minus side will generally have a better frequency performance than an NMOS transistor. Note that the previous discussion is also valid for N-type substrate technologies with PMOS and NMOS transistors exchanging roles.

As explained in MOS structures, the only problem with the accumulation condition is that the flat-band capacitance (zero potential difference across the capacitor) is less than C_{ox} and it starts to increase with voltage across the capacitor until it gets to the full C_{ox} value at around 1V, which is a voltage-dependent behavior in the voltage range 0 to 1V. In order to reduce this voltage-dependent behavior for that range, another device was introduced specially to implement capacitors in CMOS technologies, which is the well-capacitor. The well capacitor is the most commonly used capacitor in CMOS technology. It is simply a MOS structure, where the top plate is implemented using the polysilicon layer, and the bottom plate is implemented using an n^+ well in a P-substrate technology, or a p^+ well in an N-substrate technology. Figure 4-14 shows a well capacitor built on a P-substrate technology. The well-capacitor in this case can be viewed as a PMOS transistor without source and drain, and with an n^+ bulk instead of a lightly doped n-type bulk. Therefore, the same restrictions mentioned about using PMOS transistors as capacitors also apply for well-capacitor with the exception that the highly doped well in the well-capacitors has the effect of increasing the flat-band

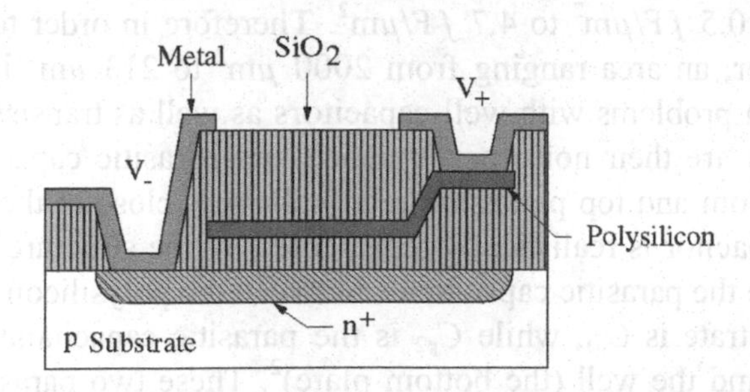

Figure 4-14. A well-capacitor on P-substrate technology.

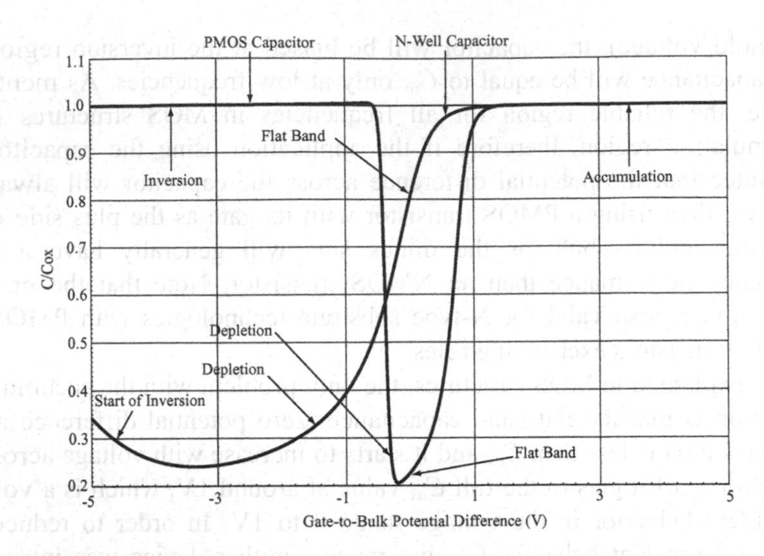

Figure 4-15. A comparison between the C-V characteristics of a PMOS transistor used as a capacitor and a well-capacitor.

capacitance as predicted by Eq. 4-9. This effectively shifts the C-V curve to the left, and improves the constant-capacitance voltage range. Figure 4-15 shows a comparison between the C-V curves at low frequencies of a PMOS transistor used as a capacitor and an N-type well-capacitor in a P-substrate technology. Note that in the well-capacitor case the threshold voltage is sufficiently high (due to the higher doping of the well) that for practical negative voltages the well-capacitor will always be in the depletion region.

In modern technologies the specific capacitance for well-capacitors ranges from 0.5 $fF/\mu m^2$ to 4.7 $fF/\mu m^2$. Therefore in order to implement a 1pF capacitor, an area ranging from 2000 μm^2 to 213 μm^2 is needed. The two common problems with well-capacitors as well as transistors connected as capacitors are their noise performance and parasitic capacitances. Since both the bottom and top plates are relatively very close to the substrate, the resulting capacitor is realistically represented by the structure shown in Fig. 4-16a, where the parasitic capacitance between the polysilicon (the top plate) and the substrate is C_{p1}, while C_{p2} is the parasitic capacitance between the polysilicon and the well (the bottom plate)[2]. These two parasitic capacitors affect the accuracy of the intended value of the capacitor. Moreover, noise from the substrate can be injected to the main capacitor through the parasitic capacitors, which degrades the noise performance of the main capacitor. One way to avoid the noise injection is using a guard-ring around the capacitor to absorb the noise from the substrate as shown in Fig. 4-16b.

In some processes, there are two available polysilicon layers that could also be used to implement capacitors. In that case, the bottom plate is implemented using the first polysilicon layer and the top plate is implemented using the second polysilicon layer. A poly-poly capacitor is shown in Fig. 4-17. Generally a poly-poly capacitor has less parasitic capacitance since the top plate is farther away from the substrate. Moreover, in order to reduce noise injected from the substrate, a shielding well biased at a quite voltage could be used to isolate the capacitor from the substrate. Therefore, poly-poly capacitors have generally better noise performance than transistor and well-capacitors. Another major advantage of poly-poly capacitors over well-capacitors is that they have relatively less voltage-dependent and frequency-dependent behavior. The fundamental reason behind that is since both the top and the bottom plates are implemented with the highly doped polysilicon layer, accumulation, depletion, and inversion have much less effect on the capacitance value. In fact, the value of the

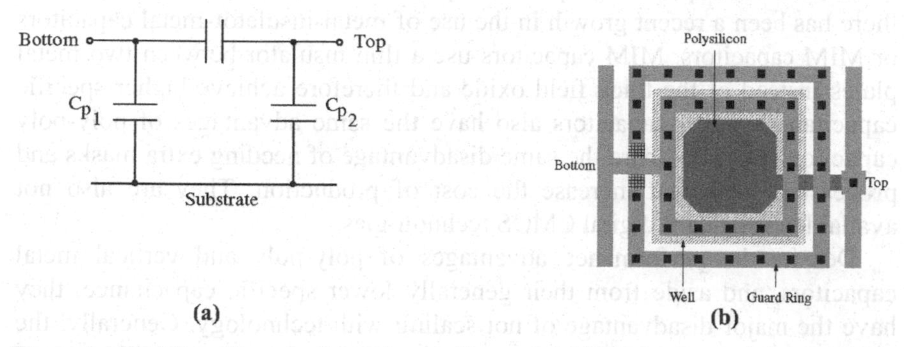

Figure 4-16. (a) Parasitic capacitance in a well capacitor, (b) Using a guard ring for noise protection.

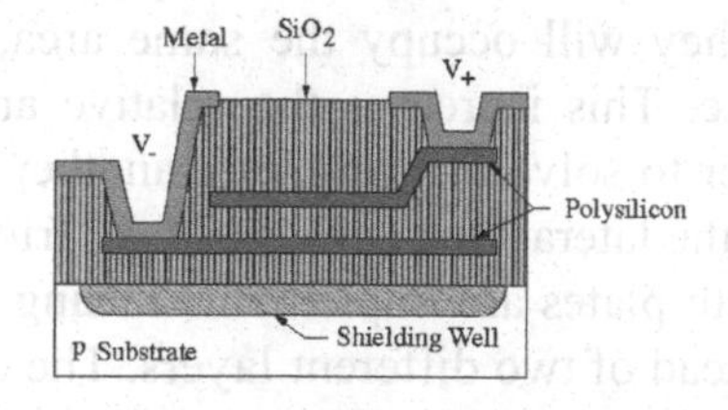

Figure 4-17. A Poly-poly capacitor.

capacitor could be found using the simple parallel plate capacitor formula in Eq. 4-8. However, poly-poly capacitors require a second polysilicon layer which adds to the cost of production due to the extra mask needed. Most digital CMOS processes doe not have a second polysilicon layer.

Another type of capacitors is the metal capacitor (aka vertical flux metal capacitor). Those capacitors use two metal layers to implement the top and bottom plates. Since metal layers are isolated from the substrate with thick field oxide, parasitic capacitance from the top and bottom plates to the substrate are very small. Therefore this type of capacitors has good noise performance. Furthermore, since metal layers are highly conductive, metal capacitors don't exhibit any voltage or frequency dependent behavior as opposed to MOS capacitors. However, since metal layers are isolated using thick field oxide, specific capacitance of metal capacitors is very low. Therefore, it is almost impractical, from an area standpoint, to use them to implement any significant capacitance.

In order to improve the specific capacitance of vertical metal capacitors there has been a recent growth in the use of metal-insulator-metal capacitors or MIM capacitors. MIM capacitors use a thin insulator between two metal plates instead of the thick field oxide and therefore achieve higher specific capacitance[6]. MIM capacitors also have the same advantages of poly-poly capacitors, but also share the same disadvantage of needing extra masks and processing steps that increase the cost of production. They are also not available in standard digital CMOS technologies.

Despite the performance advantages of poly-poly and vertical metal capacitors, and aside from their generally lower specific capacitance, they have the major disadvantage of not scaling with technology. Generally, the oxide thickness between different layers of metals and polysilicon stay almost the same even with the lateral scaling of transistors[7]. This simply means that the specific capacitance of those capacitors stays the same, which in turn means that they will occupy the same area, while the rest of the devices shrink in size. This increases the relative area of those capacitors with scaling. In order to solve this problem, another type of capacitors was introduced, which is the lateral flux capacitor (aka fringing capacitor). In this type of capacitors both plates are implemented using two metal strips of the same metal layer instead of two different layers. The capacitance is achieved through the fringing field between the two metal strips, and therefore the separation between the two metal strips is t in Eq. 4-8 and the area of the capacitor is the side area of the metal strip. Since the separation between metal lines on the same metal level scale down with the process, the specific capacitance of lateral flux capacitors increases, which in turn means that in order to get the same total capacitance after scaling, the side area of the metal strips have to also scale down with process. Figure 4-18 shows a lateral flux capacitor and how the capacitor scale with technology. Note that the different shading is just to differentiate between the two capacitor plates.

In addition to scaling with technology, lateral flux capacitors can also be incorporated with vertical capacitors to achieve a significantly higher specific capacitance[7]. As shown in Fig. 4-19, dividing a conventional vertical flux capacitor to multiple cross connected sections has the effect of increasing the total capacitance due to the lateral flux capacitors added to the structure, while keeping the total area almost the same. Effectively, this means higher specific capacitance. Note that the top and bottom plates of the equivalent capacitor are distributed between two metal layers. Different shadings are used in Fig. 4-19 to emphasize this fact. The same idea could also be extended not only to two metal layers, but to even multiple metal layers to increase the specific capacitance even further.

Since lateral flux capacitors depend on the side area of metal strips, increasing the periphery of the structure for the same area will increase the effective specific capacitance even further. There are many ways to increase the periphery, one way is shown in Fig. 4-20, where multiple metal fingers are interleaved together[8,9]. This structure is implemented in multiple metal layers and all resulting capacitors are connected in parallel. This technique

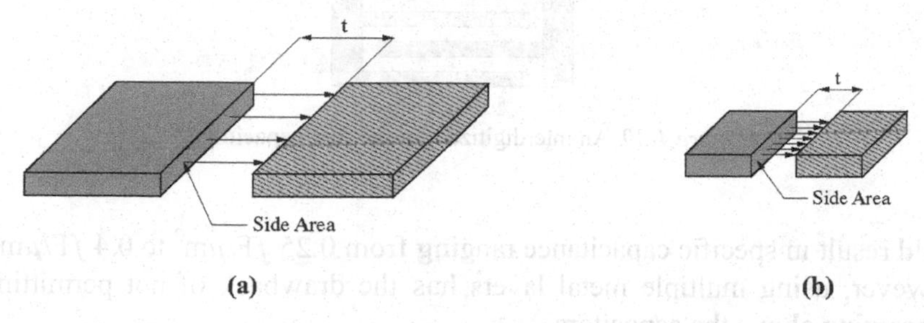

Figure 4-18. Effect of scaling on lateral flux capacitors: (a) Before scaling: lower specific capacitance and higher side area, (b) After scaling: Higher specific capacitance and lower side area.

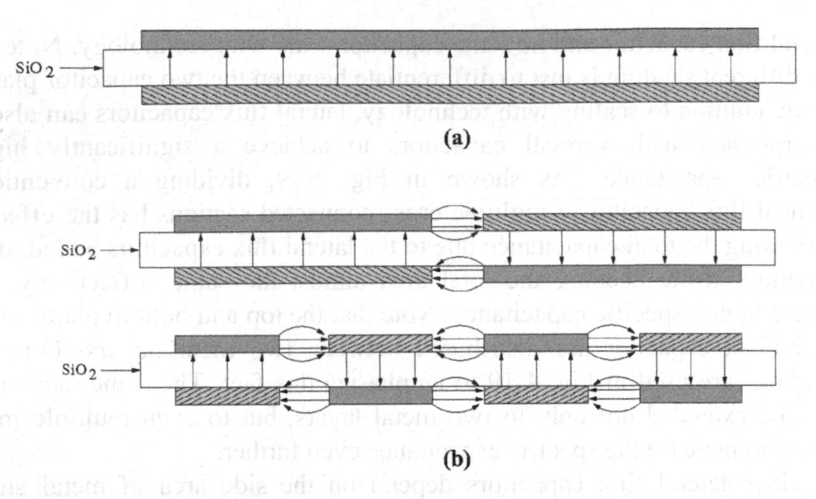

Figure 4-19. Vertical and lateral flux capacitors: (a) Conventional vertical flux capacitor, (b) Lateral flux capacitor incorporated within a vertical flux capacitor.

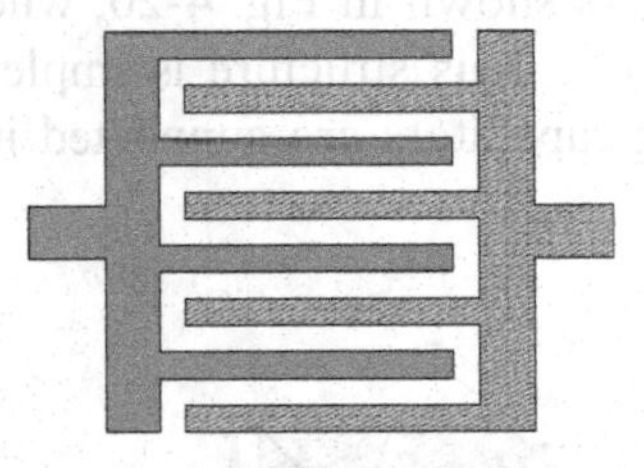

Figure 4-20. An interdigitized lateral flux capacitor.

could result in specific capacitance ranging from 0.25 $f\mathrm{F}/\mu\mathrm{m}^2$ to 0.4 $f\mathrm{F}/\mu\mathrm{m}^2$. However, using multiple metal layers has the drawback of not permitting any routing above the capacitors.

4.2.1 Accuracy of Passive Capacitors

As Eq. 4-8 suggests, the error in the value of C is a function of the error in t, L, and W. The error in the oxide thickness t is mainly a function of the accuracy of the fabrication process and there is nothing from design or layout standpoint that could minimize its impact. The error in W and L on the other hand is due to the etching process as discussed earlier in the resistor's case. Taking the error in W and L into account, and assuming that $\Delta W = \Delta L = 2x$, the actual area of the capacitor is[2]:

$$A_{\text{actual}} = (W - 2x)(L - 2x) \approx WL - 2(W + L)x = A - Px = A\left(1 - \frac{P}{A}x\right) \quad (4\text{-}12)$$

where A is the drawn area of the capacitor, and P is the perimeter. The relative error in the area of the capacitor becomes:

$$\frac{A_{\text{actual}} - A}{A} = -\frac{P}{A}x = -2\left(\frac{x}{W} + \frac{x}{L}\right) \quad (4\text{-}13)$$

As shown in Eq. 4-13, in order to reduce the effect of the error x, W and L has to be large. Unlike the resistor case, it is not possible to increase W and L without increasing the value of the capacitor. Therefore, if a lower capacitance is needed, the dimensions of the capacitor have to be smaller leading to higher errors. Adding dummies around the capacitors also helps reducing the errors in W and L due to the etching process as discussed in resistors case.

With process variations, well-capacitors vary around ±10% from their targeted value, which is generally less than the variations in resistors. The reason behind that is that the error in the resistor value is also a direct function of the error in doping in addition to the error in W and L, while doping in case of well-capacitors has a little effect on the capacitance since the well is already highly doped. This minimizes the impact of doping errors on well-capacitors, while those errors are significant in resistors.

Metal capacitors, especially interdigitized lateral flux capacitors, on the other hand vary widely with process, roughly around ±25%. This is essentially due to the fact that flux capacitors are composed of many fingers interleaved together, with each finger having its own W, L, and t errors. In addition to that and in order to increase the specific capacitance, multiple flux capacitors on multiple metal layers are usually connected together in parallel, and each metal layer introduces its own error as well. All those errors can add up to a significant total error in the value of the capacitor.

Generally capacitors do not vary significantly with temperature since errors in W, L, and t are not temperature dependent. Furthermore, carrier concentration in both the top and bottom plates of capacitors has little effect on the capacitance value due to their highly conductive nature. Therefore flux capacitors temperature coefficient is as low as -24 to -12 ppm, while well-capacitors have a temperature coefficient varying from 120 ppm to180 ppm. Well-capacitors have a higher temperature coefficient since carrier concentration in the well changes with temperature, however the impact on the capacitance is still considered very low as compared to resistors.

4.2.2 Matching Properties of Passive Capacitors

As discussed in the matching properties of integrated resistors, capacitors could also match very well given a careful layout. This fact has driven the development of many circuit techniques that depend solely on matching between capacitors rather than their absolute values such as switch capacitor circuits. Same design and layout guidelines followed in integrated resistors to achieve good matching apply as well to integrated capacitors. Therefore using dummies to minimize the effect of etching and to keep the etching profile uniform across the whole structure is very important to achieve good matching. As mentioned in the resistors case, using rounded or 45 degrees edges for the corners of the capacitor helps in minimizing etching errors.

All different levels of layout complexity in order to achieve good matching between two resistors also apply to capacitors. Figure 4-21 shows two matched capacitors laid out in a common-centroid fashion. Note how the contacts between the polysilicon top plates and the metal layer are done on top of the thin oxide, which is permissible in some processes. In some other processes, those contacts have to be done on top of the field oxide as shown in Figs. 4-14 and 4-16b. Very often, a specific ratio between two capacitors is needed rather than just two matched capacitors. Since the ratio between two capacitors is essentially equal to the ratio between their respective areas, Eq. 4-12 suggests that if the two capacitors had the same perimeter to area ratio, the effect of the etching error x on matching could be eliminated. Hence, it is very crucial to keep P/A the same for the two capacitors. One way to do that is to use a unit capacitor to build both capacitors as multiple parallel unit capacitors. Figure 4-22a shows two matched capacitors with a ratio 1:2 built of a unit capacitor. This method is useful if the ratio needed is a ratio between integers. If the ratio between the two capacitors is not a ratio of integers, then a unit capacitor is used to implement the integer part of the ratio, and a non unit capacitor with the same P/A is used to implement the fractional part of the ratio. Figure 4-22b shows the layout of two matched capacitors with a ratio 4:3.2.

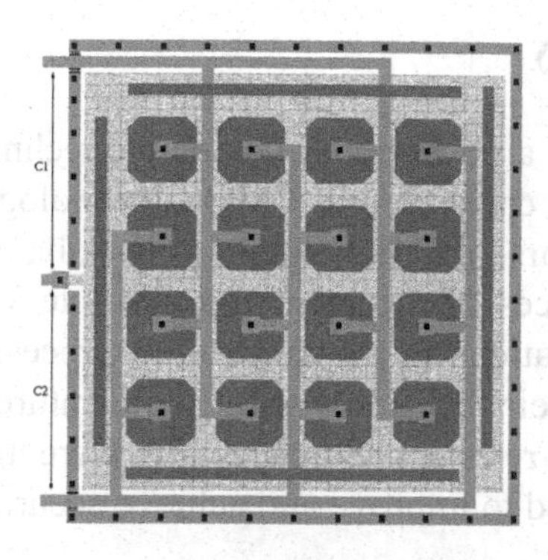

Figure 4-21. Layout of two matched capacitor.

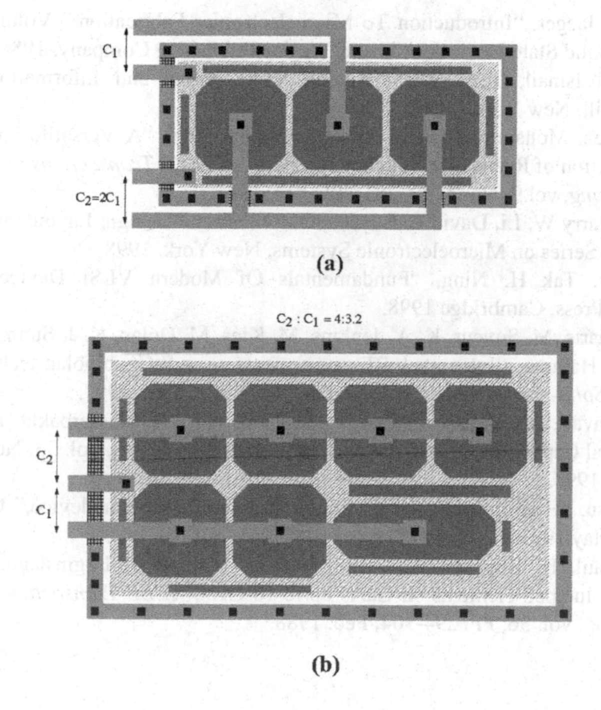

Figure 4-22. (a) Layout of an integer multiple of a unit capacitor, (b) Layout of a non integer multiple of a unit capacitor.

4.3 Summary

This chapter gave an overview of available techniques for implementing on-chip resistors and capacitors in CMOS technologies. This includes poly and diffusion resistors and poly-well, poly-poly, and lateral flux metal capacitors. The sources of errors in the absolute values of those on-chip elements due to variations in the fabrication process were presented along with a study of their behavior with temperature variations. Matching properties of on-chip resistors and capacitors were also presented as well as layout techniques used to improve and achieve accurate matching.

LIST OF REFERNCES

1. Richard C. Jaeger, "Introduction To Microelectronics Fabrication," Volume V, Modular Series On Solid State Devices, Addison-Wesley Publishing Company, 1988.
2. Mohammed Ismail, Terri Fiez, "Analog VLSI Signal and Information Processing," McGraw-Hill, New York, 1994.
3. Frode Larsen, Mohammed Ismail, and Christopher Abel, "A Versatile Structure for On-Chip Extraction of Resistance Matching properties," *IEEE Transactions on Semiconductor Manufacturing*, vol.9, No. 2, PP. 281-285, Feb. 1988.
4. R. Jacob, Harry W. Li, David E. Boyce, "CMOS Circuit Design, Layout, and Simulation," IEEE Press Series on Microelectronic Systems, New York, 1998.
5. Yuan Taur, Tak H. Ning, "Fundamentals Of Modern VLSI Devices," Cambridge University Press, Cambridge 1998.
6. J. N. Burgharts, M. Soyeur, K. A. Jenkins, M. Kies, M. Dolan, K. J. Stein, J. Malinowski, and D. L. Harame, "Integrated RF components in a SiGe bipolar technology," *IEEE Journal of Solid-State Circuits*, vol. 32, PP. 1440-1445, Sept. 1997.
7. Hirad Samavati, Ali Hajimiri, Arvin R. Shahani, Gitty N. Nasserbakht, and Thomas H. Lee, "Fractal Capacitors," *IEEE Journal of Solid-State Circuits*, vol. 33, No. 12, PP. 2035-2041, Dec. 1998.
8. O.E. Akcasu, "High capacitance structures in a semiconductor device," U.S. Patent No. 5208725, May 1993.
9. E. Pettenpaul, H. Kapusta, A. Weisgerber, H. Mampe, J. Luginsland, and I. Wolff, "Models of lumped elements on GaAs up to 18 GHz," *IEEE Transactions on Microwave Theory Tech.*, vol. 36, PP. 294-304, Feb. 1988.

Chapter 5

A DIGITAL ADAPTIVE TECHNIQUE FOR ON-CHIP RESISTORS

As outlined in chapter 4, on-chip resistors are a very attractive solution for implementing systems that require resistors due to cheaper cost and higher integration levels associated with them. However, the limited control over their absolute values, especially in digital CMOS processes, could potentially degrade the performance of the whole system. Most processes guarantee the absolute values of on-chip resistors with only ±25% accuracy, which has been a major obstacle for fully integrating systems that require highly accurate resistors[1].

Many techniques were developed to implement accurate on-chip resistors using automatic tuning techniques[2-5]. Some techniques use tunable, purely-active devices (transistors) configured to cancel any nonlinearity in transistors behavior[6-12]. Those transistors are placed in a feedback loop in order to maintain their effective resistance to an accurate level. Usually though, since those transistors constitute 100% of the total resistance, the accuracy of their gate-to-source voltage is crucial for determining their equivalent resistance, which implies that an on-chip accurate voltage reference (a band-gap) needs to be implemented and used as a reference to generate the gate-to-source voltage. Furthermore, since the transistors are used as analog elements (not as digital switches), the tuning process that provides their gate-to-source voltage has to be performed in the analog domain, which has multiple disadvantages. First, there is no simple, precise, and power-efficient way to store analog values, therefore, the analog tuning loop that provides the tuning voltage to the transistors (usually their gate-to-source voltage) has to either be operating at all times, or the result of the tuning process has to be stored on a capacitor and periodically refreshed. Both alternatives are power and area inefficient. Second, analog tuning is generally more susceptible to noise especially when the tuned resistance is implemented using only transistors. This implies that any noise glitches on the tuning voltage will have direct impact on the value of the implemented resistance. Third, analog tuning does not generally provide a simple, area-efficient way to tune several and different-valued on-chip resistors independently. The reason behind that is that it is generally very difficult to design different-valued resistors (implemented using transistors) such that they all use the same tuning voltage range, especially if the ratio between those resistors is fractional. Thus, it is very difficult to use the same tuning loop to independently tune all the different resistors on the chip, and in most cases each resistance requires a separate tuning loop, which is an area and power-expensive solution. In addition to the disadvantages of analog tuning mentioned above, active devices with nonlinearity cancellation techniques still generally suffer from range, linearity, and noise problems. The range problem evolves from the fact that active transistors need a specific minimum voltage to be turned on, as well as any other range restriction to keep them in a specific operation mode (triode, or saturation), while the linearity problem evolves from the fact that nonlinearity cancellation techniques mostly rely on the classic square-law representation of transistors, which has not been very accurate in modeling nonlinearities in digital CMOS processes[13]. On the other hand, since active elements are in direct contact with the substrate, they are relatively more vulnerable to substrate noise coupling. This limits noise performance of the implemented resistor, especially when active devices constitute 100% of the resistor.

Some other techniques use a combination of on-chip passive resistors and active devices together to implement the required resistance[14]. In that case, the active device is usually used to correct for errors in an on-chip passive resistor (usually by connecting the active device in parallel with the on-chip passive resistor). Since the correction is still achieved through the resistive contribution of an active device, the total resistance still suffers from range and linearity limitations (although less than purely-active implementations). Additionally, an accurate voltage reference and analog tuning loops are still needed. Switched-C techniques are also usually used in the area of integrated filters to implement on-chip resistance using ratio between capacitors, but they require clocks and can only operate in discrete-time domain[15].

This chapter describes a digital tuning algorithm used to implement accurate on-chip resistances using passive elements[16,17]. The technique has the advantage of using a single off-chip resistor to tune all required on-chip resistors automatically even if they are of different values or scattered around different areas of the chip. The algorithm achieves a tight control over on-chip resistors and is limited only by the accuracy of the off-chip resistor, mismatches, and area restrictions. The technique does not require any accurate voltage references and it implements the required resistance using on-chip passive resistors. Furthermore, the technique limits the use of active devices to digital switching purposes with a negligible contribution to the total resistance and without having to design those switches with large dimensions. This fact reduces the noise introduced to the system and maintains high linearity, wide range of operation, and high speed performance. After discussing the details of the technique, application example is presented, that is, transmission line terminations for two high-speed wire line transceivers (480 Mbps, and 1.65 Gbps) that require ±10% accuracy.

5.1 The Calibration Technique

Implementing accurate on-chip resistors using active elements with nonlinearity cancellation techniques, or using well or polysilicon resistors requires essentially some sort of an automatic tuning procedure. The automatic tuning process has three fundamental steps. The first step is the design of an electronically tunable element (a tunable resistor in this case). The second step is quantifying the drift in the parameter of interest (which is the resistance value in this case). The third step is generating a tuning signal based on the detected drift in the parameter of interest to adjust it back to its desired value. Therefore, a tunable resistor, a method for quantifying the drift in the resistance value, and a tuning signal generation method need to be developed in order to perform the tuning process.

Most modern mixed-signal ICs essentially have an off-chip resistor and a band-gap reference circuit, where both are used to generate accurate biasing currents and voltages needed by analog parts of the circuit. The calibration architecture uses this same off-chip resistor as a reference without interfering with the bias-current generation process. However, the calibration technique does not need the band-gap reference or any kind of an accurate voltage reference. Before the technique is introduced, a discussion of the concept of "variation range quantization" is presented.

5.1.1 Variation Range Quantization Concept

The concept of "variation range quantization" is shown in Fig. 5-1. Let's assume that an on-chip resistor is designed to have the value R_{req} in nominal conditions. With process and temperature variations, this value will drift. Let's define the maximum percentage an on-chip resistor can drift from its nominal value R_{req} to be $\pm D_{Max}\%$, where D_{Max} is a characteristic of the fabrication process, physical dimensions of the resistor, and temperature. It is usually available through process characterization. Now, let's define the maximum percentage error allowed in the on-chip resistor from its nominal value R_{req} to be $\pm C_{Max}$%. As shown in Fig. 5-1, the whole $\pm D_{Max}$ range can be divided into $(n+1)$ different regions, where each region is C_{Max} wide except for region $(n/2)$ which is $2C_{Max}$ wide. Region $(n/2)$ is essentially where the tuning algorithm needs to control the variation in the on-chip resistance to be within. This leads to the following equation that determines n:

$$n = \left(\frac{D_{Max}}{C_{Max}} - 1 \right) \times 2 \qquad (5\text{-}1)$$

Therefore, what the tuning algorithm should achieve is to always tune the variation in the resistor value to region $(n/2)$ if the variation in the resistor value falls within any of the different $(n+1)$ regions in Fig. 5-1.

5.1.2 The Resistor Block

Given the explanation of the "variation range quantization" concept presented earlier, a logical starting point would be implementing the required resistance using a resistor block that is composed of $(n+1)$ parallel on-chip resistors as shown in Fig. 5-2 instead of a single resistor, where each resistor is controlled with a MOS switch that can be turned on or off using a digital control signal. In Fig. 5-2, R_0 constitutes the core resistance and its value is normally close to R_{req}, while the rest of the resistors are just modulating resistors that could be switched on or off to adjust the total resistance of the resistance block to be within $R_{req}\pm C_{Max}$%. For example, if in nominal conditions the parallel combination of R_0 up to $R_{n/2}$ is designed to be exactly equal to R_{req}, then errors due to process variations could be corrected as follows: If the drift in the resistors values due to process variations is negative (lower resistors values), then the error can be corrected by switching off some of the parallel resistors R_1 up to $R_{n/2}$ depending on the amount of the drift. On the other hand, if the drift is positive (higher resistors values), then the error could be corrected by switching on some of the parallel resistors $R_{(n/2)+1}$ up to R_n. Therefore, given the amount of error in the value of the on-chip resistors that compose the resistor block, the digital control signals should assume a digital value to counteract this variation. This resistor block represents the electronically tunable element in the proposed tuning architecture. The next section presents a mathematical way for determining the number of resistors in the resistor block, the value of each resistor, and how to develop the control signals.

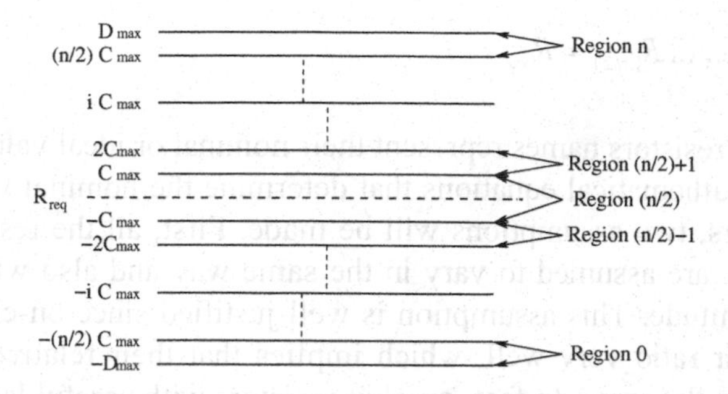

Figure 5-1. The "variation range quantization" concept. The error range in the value of on-chip resistors is being divided to *n* equal regions, with each region equal to the maximum targeted error.

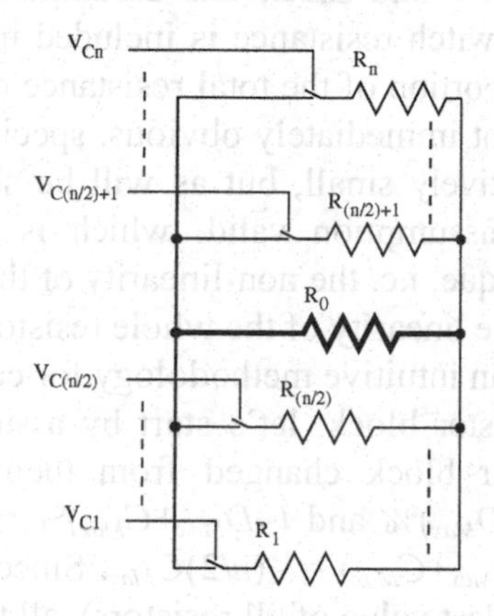

Figure 5-2. The resistor block.

5.1.3 The Algorithm

To start off, if each variation region in Fig. 5-1 is assigned to one of the parallel resistors in the resistor block, one can find adequate values for those resistors to guarantee that the total resistance of the resistor block is within region (n/2). As discussed earlier, the resistor is designed such that in nominal condition R_1 through $R_{(n/2)}$ are switched on, while $R_{(n/2)+1}$ through R_n are switched off. Note that R_0 is always connected with no switches, which is an advantage that will be discussed later in details. Based on the errors in the values of those resistors, the tuning algorithm should either switch on or switch off more resistors in parallel to R_0 in order to always place the total resistance of the resistor block into region (n/2). The previous discussion implies that:

$$R_0 // R_1 // ....R_{(n/2)} = R_{req} \tag{5-2}$$

Note that the resistors names represent their nominal or ideal values. In order to develop mathematical equations that determine the nominal values of the (n+1) resistors, two assumptions will be made. First, all the resistors in the resistor block are assumed to vary in the same way and also with the same relative magnitude. This assumption is well justified since on-chip resistors maintain their ratio very well, which implies that their relative magnitude change is also the same. In fact, on-chip resistors with careful layout usually match within better than 0.1% given that all of them are placed in close proximity and also have close values[1]. Provided that those resistors are laid out carefully, and are designed to have wide dimensions to minimize under etching effects, i.e. $\Delta W/W$ and $\Delta L/L$, this assumption is safe. The second assumption is that the switch resistance is included into the resistors values and constitutes a small portion of the total resistance of each resistor. This is an assumption that is not immediately obvious, specially when the required resistance value is relatively small, but as will be shown later, the tuning algorithm makes this assumption valid, which is one of the important advantage of this technique, i.e. the non-linearity of the switch resistance has a very small effect on the linearity of the whole resistor block.

In order to develop an intuitive methodology for calculating the values of the resistors in the resistor block, let's start by assuming that the on-chip resistors in the resistor block changed from their nominal value by a percentage between $(-D_{Max})$% and $(-D_{Max}+C_{Max})$%, which is region (0) in Fig. 5-1. Note that $(-D_{Max}+C_{Max}) = -(n/2)C_{Max}$. Since this is the worst case negative error (the smallest value of all resistors), all the resistors R_1 through $R_{n/2}$ should be switched off leaving only R_0 connected. Note that by using this strategy, R_0 does not need a switch, which is a very beneficial fact as will be explained later on. Since in this case only R_0 is present, it has to satisfy the following two inequalities to cover its own errors:

$$R_0\left(100 - D_{Max}\right) > \left(100 - C_{Max}\right) R_{req} \tag{5-3}$$

$$R_0\left(100 - \frac{n}{2} C_{Max}\right) < \left(100 + C_{Max}\right) R_{req} \tag{5-4}$$

Equation 5-3 guaranties if the error is at the bottom of region (0) $(-D_{Max}\%)$, then the nominal value of R_0 will be high enough to place the total resistance of the resistor block (including the error) at a value higher than the bottom edge of region (n/2). Equation 5-4 on the other hand guaranties that if the error is at the top of region (0) [($-D_{Max}+C_{Max}$)%)], then the nominal value of R_0 will be low enough not to place the total resistance of the resistor block (including the error) at a value higher than the top edge of region ($n/2$). Both equations guarantee that for any error within ($-D_{Max}$)% and ($-D_{Max}+C_{Max}$)%, the total resistance of the resistor block (including the error) will always be within $\pm C_{Max}$% from the required nominal value R_{req}. Combining Eqs. 5-3 and 5-4 into a single inequality:

$$\frac{100-C_{Max}}{100-D_{Max}}R_{req} < R_0 < \frac{100+C_{Max}}{100-\frac{n}{2}C_{Max}}R_{req} \tag{5-5}$$

Equation 5-5 determines the range of values that R_0 can nominally take. Now, assuming that the resistors in the resistor block changed some where between ($-D_{Max}+C_{Max}$)% and ($-D_{Max}+2C_{Max}$)% from their nominal value, and assuming in this case that both R_0 and R_1 are switched on, then the same way R_0 was determined, the following inequality can be used to determine R_1:

$$\frac{100-C_{Max}}{100-\frac{n}{2}C_{Max}}R_{req} < R_1 // R_0 < \frac{100+C_{Max}}{100-\frac{n-2}{2}C_{Max}}R_{req} \tag{5-6}$$

solving Eq. 5- 6 for R_1:

$$\frac{R_0\left(100-C_{Max}\right)R_{req}}{100\left(R_0-R_{req}\right)+C_{Max}\left(R_{req}-\frac{n}{2}R_0\right)} < R_1$$

$$R_1 < \frac{R_0\left(100+C_{Max}\right)R_{req}}{100\left(R_0-R_{req}\right)-C_{Max}\left(R_{req}+\frac{n-2}{2}R_0\right)} \tag{5-7}$$

Again, Eq. 5-7 defines a range of values the nominal value of R_1 can take as a function of R_0, while R_0 could be determined using Eq. 5-5. If the same method is followed (adding more resistors in parallel) to calculate the rest of the resistors, the following set of recursive inequalities can be determined:

$$\frac{(100-C_{Max})R_{req}\,RT_i}{100\left(RT_i-R_{req}\right)+C_{Max}\left(R_{req}-\left(\frac{n}{2}+1-i\right)RT_i\right)}<R_i$$

$$R_i<\frac{(100+C_{Max})R_{req}\,RT_i}{100\left(RT_i-R_{req}\right)-C_{Max}\left(R_{req}+\left(\frac{n}{2}-i\right)RT_i\right)} \tag{5-8}$$

for $i < n/2$, and

$$\frac{(100-C_{Max})R_{req}\,RT_i}{100\left(RT_i-R_{req}\right)+C_{Max}\left(R_{req}-\left(\frac{n}{2}-i\right)RT_i\right)}<R_i$$

$$R_i<\frac{(100+C_{Max})R_{req}\,RT_i}{100\left(RT_i-R_{req}\right)-C_{Max}\left(R_{req}+\left(\frac{n}{2}-i-1\right)RT_i\right)} \tag{5-9}$$

for $i > n/2$, and

$$\frac{1}{R_{\frac{n}{2}}}=\frac{1}{R_{req}}-\frac{1}{RT_{\frac{n}{2}}} \qquad \text{for } i = n/2 \tag{5-10}$$

where i is an index that starts from 0 to n in order to calculate resistors R_0 up to R_n, and RT_i is the total parallel combination of R_0 up to R_{i-1}, or mathematically:

$$\frac{1}{RT_i}=\sum_{k=0}^{k=i-1}\frac{1}{R_k} \quad \text{for } i>1 \text{ and } \quad RT_0=\infty \quad \text{for } i>0 \tag{5-11}$$

Note that for the special case when $i = 0$ and $RT_0 = \infty$, Eq. 5-8 reduces to Eq. 5-5. Note also that Eq. 5-10 is another form of Eq. 5-2.

5.1.4 Quantifying the Drift

After the mathematical equations that determine the nominal values of the resistors in the resistor block have been developed (which represent the first step of the tuning process), a technique to quantify the drift in the resistance value has to be developed. This drift-quantification process represents the second step of the tuning process. In order to achieve that, the circuit in Fig. 5-3 is used. The op-amp copies a reference voltage V_B to an off-chip resistor R_{ext}. The current generated is then mirrored and injected to an on-chip resistor R_{int}. Note that the simple current mirror shown in Fig. 5-3 is only conceptual, a high output impedance current mirror designed to minimize the effects of transistor mismatches on the mirroring ratio K should be used. Usually, a cascode or a low-voltage cascode current mirror is sufficient. The voltage across R_{int} (V_{Br}) is:

$$\frac{V_{Br}}{V_B} = \frac{R_{int}}{K\,R_{ext}} \tag{5-12}$$

Equation 5-12 shows that the relative drift in R_{int} from KR_{ext} (which is why R_{ext} is considered the reference) will cause the same relative drift in V_{Br} from V_B. Therefore, if R_{int} is implemented using the same resistor type as in the resistor block and laid out in a close proximity to it, then by comparing V_{Br} to V_B, the relative error in the values of on-chip resistors in the resistor block due to process and temperature variations can be detected. It is worth mentioning that since V_{Br} is being compared to V_B to find the relative change, the absolute value of V_B or its accuracy is of no concern. In fact, it could simply be just a potential divider from the supply and it does not matter if it changes with process or temperature. It is also worth mentioning that the topology used in Fig. 5-3 is usually used to generate accurate biasing currents by tapping off more current mirrors and using a band gap circuit to provide V_B. Therefore, if the system requires those accurate currents, then the described tuning architecture could simply use it for its purposes without interfering with the bias-current generation process. The factor K as well as R_{int} could simply be anything desired, but since V_{Br} is being compared to V_B, it is much easier just to set the ratio in Eq. 5-12 to be nominally unity. This implies that the nominal value of R_{int} (its desired ideal value) will be K times R_{ext}. This gives more flexibility in choosing the value of R_{int}, for example picking R_{int} to be double R_{ext} reduces power consumption by a factor of two on the expense of doubling the area needed to implement R_{int}.

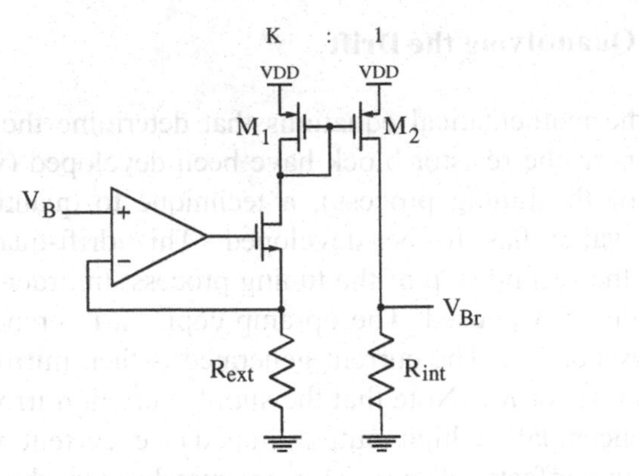

Figure 5-3. Evaluating the drift in the on-chip resistor R_{int} from the reference resistor R_{ext} by comparing V_{Br} to V_B.

5.1.5 Tuning Signal Generation

As shown by the "variation range quantization" concept discussed earlier, the state of the switches controlling the resistors in the resistor block is determined by finding the region in which the relative error in the values of the on-chip resistors from their nominal value occur. Since the ratio V_{Br}/V_B is essentially the relative change in the values of the resistors, and using the borders of each variation region shown in Fig. 5-1, then according to the algorithm presented earlier the states of the switches can be determined as follows:

If $V_{Br} > \left(1 - \frac{n}{2\times100}C_{Max}\right)V_B$, then resistor R_1 must be switched on. Let's call the digital control signal that controls the switch to be V_{C1} and let's assume that it's active high. This leads to the logical equation:

$$V_{C1} = high \text{ if } V_{Br} > V_{r1} \text{ and } V_{C1} = low \text{ if } V_{Br} < V_{r1} \tag{5-13}$$

where $V_{r1} = \left(1 - \frac{n}{2\times100}C_{Max}\right)V_B$. Thus, the digital control signal V_{C1} that controls the switch is simply the result of the comparison of V_{Br} to the reference voltage V_{r1} which corresponds to the lower border of region (1) in Fig. 5-1. Using the same steps, the following set of recursive equations that determine the control signals for all the switches as well as the reference voltages needed for the comparison process with V_{Br} could be written as:

$$V_{ri}=\left(1-\left(\frac{\frac{n}{2}+1-i}{100}\right)C_{Max}\right)V_B \qquad \text{for } i\le n/2 \qquad (5\text{-}14)$$

$$V_{ri}=\left(1-\left(\frac{\frac{n}{2}-i}{100}\right)C_{Max}\right)V_B \qquad \text{for } i> n/2 \qquad (5\text{-}15)$$

and for $i\le n/2$:

$$V_{Ci}=high \text{ if } V_{Br}>V_{ri} \text{ and } V_{Ci}=low \text{ if } V_{Br}<V_{ri} \qquad (5\text{-}16)$$

while for $i>n/2$:

$$V_{Ci}=low \text{ if } V_{Br}<V_{ri} \text{ and } V_{Ci}=high \text{ if } V_{Br}>V_{ri} \qquad (5\text{-}17)$$

where i is an index that starts from 1 to n. Note how in nominal conditions, the logic levels of V_{Ci} are inverted for $i > n/2$.

The reference voltages V_{ri} in Eqs. 5-14 and 5-15 are essential for the tuning process. As shown by those equations, those voltages are fractions of the reference voltage V_B. Thus, the reference voltage generator shown in Fig. 5-4 is used. The op-amp forces V_B across the series resistors R_{r0} to $R_{r(n/2)}$, and due to the infinite input impedance of the op-amp, the resulting current will also flow through the resistors $R_{r(n/2+1)}$ to $R_{r(n)}$. Using Eqs. 5-14 and 5-15, those resistors could be found using the following recursive equation:

$$R_{ri}=\left(1-\left(\frac{\frac{n}{2}-i}{100}\right)C_{Max}\right)R_{rt}-\sum_{k=0}^{k=i-1}R_{rk} \qquad (5\text{-}18)$$

where i is an index that starts from 0 to n, and $R_{rt}=\sum_{k=0}^{k=n/2}R_{rk}$. Note that R_{rt} along with V_B determines the amount of current flowing in the potential divider. Note also that when $i=0$, $\sum_{k=0}^{k=-1}R_{rk}=0$.

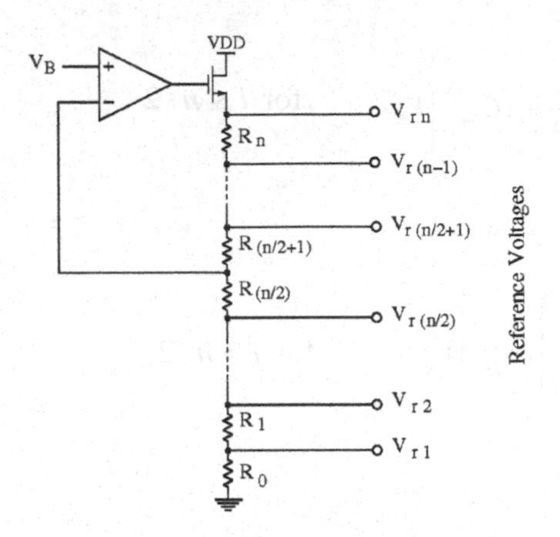

Figure 5-4. The voltage reference generator.

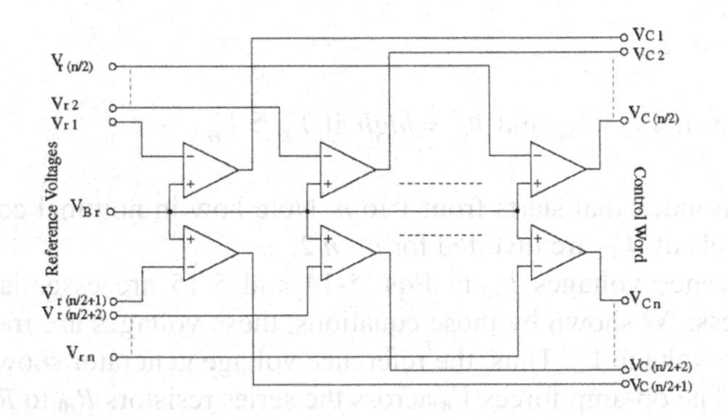

Figure 5-5. A conceptual implementation of Eqs. 5-16 and 5-17

Once the reference voltages are generated using the circuit in Fig. 5-4, generating the actual control signals represented by Eqs. 5-16 and 5-17 becomes just a matter of comparing the reference voltages with V_{Br}. The control matrix shown in Fig. 5-5 is a conceptual implementation of Eqs. 5-16 and 5-17 in which a stack of n comparators are used to perform the comparison process and generate the control signals accordingly. Note that the control matrix could be physically implemented as in Fig. 5-5, in which the whole tuning process completed within one comparator delay on the expense of using n comparators, or it could be implemented with a single comparator and a state machine, where reference voltages are successively switched to the input of the comparator and the results being stored. In this case, die area is saved on the expense of longer tuning time. Fig. 5-6 shows the complete calibration architecture.

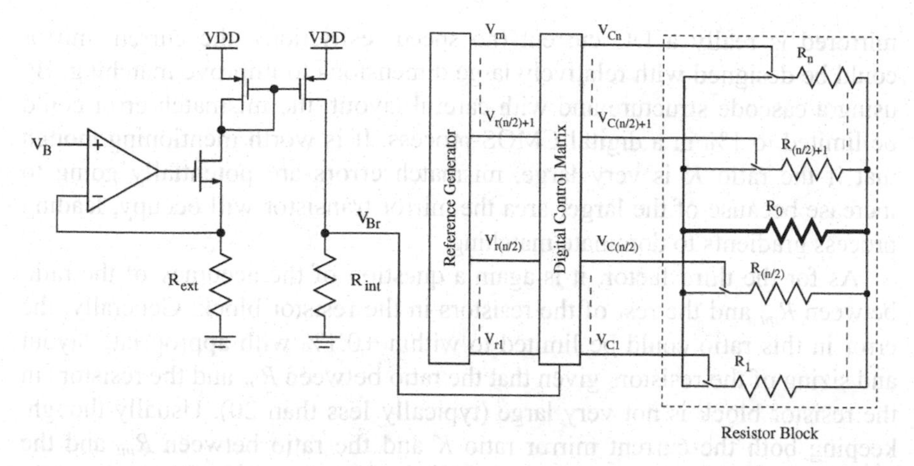

Figure 5-6. The complete calibration architecture.

5.2 Practical Advantages and Limitations

In the previous few sections, the calibration technique has been described and explained with no discussion of practical advantages and performance limitations. In the next few sections, some of those advantages as well as accuracy limitations are presented along with some important design considerations in the implementation process of the architecture.

5.2.1 Accuracy Limitations

Considering the described algorithm, the accuracy of the tuning process is limited by four factors. The first factor is the accuracy of the reference off-chip resistor R_{ext}. The second factor is the ratio V_{Br}/V_B and how accurate it really represents the drift of R_{int} from the reference R_{ext}. The third factor is how close the relative error detected in the value of R_{int} (through the ratio V_{Br}/V_B) is to the actual relative error in the values of the resistors in the resistor block. The fourth factor is the accuracy of the comparison process between V_{Br} and the reference voltages shown in Fig. 5-4. As will be discussed later, the second and third factors are really coupled to each other.

As for the first factor, typical off-chip resistors have an accuracy of ±1%, but they are also available with an accuracy of up to ±0.1% for a significantly higher cost. The accuracy of the off-chip resistor represents the maximum theoretical accuracy of the algorithm. As for the second factor, the accuracy of the current mirror ratio K in Eq. 5-12 defines how accurate the ratio V_{Br}/V_B really represents the drift of R_{int} from R_{ext}. Therefore, it becomes a question of mismatch errors in the current mirror. Since the current being mirrored is really a DC current (no speed restriction), the current mirror could be designed with relatively large dimensions to improve matching. By using a cascode structure and with careful layout, the mismatch error could be limited to 1% in a digital CMOS process. It is worth mentioning though that if the ratio K is very large, mismatch errors are potentially going to increase because of the larger area the mirror transistor will occupy, leading process gradients to dominate matching.

As for the third factor, it is again a question of the accuracy of the ratio between R_{int} and the rest of the resistors in the resistor block. Generally, the error in this ratio could be limited to within ±0.1% with appropriate layout and sizing of the resistors given that the ratio between R_{int} and the resistors in the resistor block is not very large (typically less than 20). Usually though, keeping both the current mirror ratio K and the ratio between R_{int} and the resistors in the resistor block low is not necessarily simultaneously achievable, especially if the required resistance is small (50Ω for example). In this case a compromise between these two accuracy limitations has to be reached. In general terms, the mismatch error in current mirrors is much more significant with a larger mirroring ratio than the mismatch error between resistors with the same large ratio (provided that the resistors are poly resistors). Hence, it is always better from an accuracy standpoint to keep the mirroring ratio K small and accept a large ratio between R_{int} and the resistors in the resistor block. Having said that, the ±0.1% error in the ratio between the resistors mentioned earlier could be no longer achievable and a higher error should be expected and accounted for. In order to improve that error though, R_{int} and the rest of the resistors in the resistor block could be implemented using parallel and series combinations of a unit resistor. Ideally, the unit resistance used should be the geometric mean of the largest and smallest resistors, but this implies that all the resistors have to be an integer multiple of that unit resistor, which is not possible since the values of the resistors are set by the algorithm. Therefore, using the unit resistor will imply that some of the resistors will have to use fractional resistors in addition to integer multiple of the unit resistor. To improve matching even further, it's best for the unit resistor not to have minimum dimensions, however, this has a serious area penalty.

As for the fourth factor, there are two elements that determine the accuracy of the comparison process. First, the accuracy of the ratio between the reference voltages shown in Fig. 5-4 and V_B, and how well those ratios really represent the borders of the variation regions shown in Fig. 5-1. Second, the error in the comparison process between V_{Br} and the reference voltages due to the input offsets of the comparators. As for the first element, it's again a question of matching between the resistors that compose the circuit shown in Fig. 5-4. With careful layout and sizing of those resistors, mismatch error could be limited to ±0.1%[18]. As for the second element, the input offset of the comparators will have no effect on the results of the comparison process unless the ratio V_{Br}/V_B is within the input offset of the comparator from the border lines of the variation regions in Fig.1. In this case, the error introduced will depend on the ratio between the input offset voltage of the comparator and the reference voltage V_B. The input offset voltage could be easily limited to less than 5mV with careful layout, and if V_B is chosen to be 1.2V, the error could be limited to ±0.4%.

From the previous discussion, the achievable accuracy of the algorithm is limited to the accuracy of the off-chip resistor and mismatch errors between similar on-chip elements. So in essence, the algorithm makes the accuracy of on-chip resistors a question of matching between similar on-chip elements instead of a question of absolute values accuracy. The same concept used to replace resistors in switched capacitors circuits. The major drawback of the technique though is the relatively larger area required in order to achieve good matching between R_{int} and the different resistors in the resistor block.

5.2.2 Practical Advantages

The described technique has multiple practical advantages that are worth pointing out. The first advantage is the recursive nature of Eqs. 5-8 to 5-10 and 5-14 to 5-18. This recursive nature makes the design process very simple. An excel sheet or a matlab code could simply be developed to solve those equations. The required resistance value R_{req}, the maximum variation D_{Max}, the required tolerance C_{Max}, and R_{rt} could be introduced to the excel sheet or matlab code as inputs, and all the resistors in the resistor block as well as the reference generator could be calculated.

The second advantage is that Eqs. 5-8 to 5-10 give a range of values that each resistor in the resistor block can take rather than a specific accurate value. This makes the design very flexible to inaccuracies and very easy to implement. Normally, the value of each resistor in the resistor block should be in the middle of the range specified by Eqs. 5-8 to 5-10 in order to achieve maximum margin possible.

The third advantage is that the modulating resistors R_1 through R_n are significantly higher than the core resistor R_0. For example, if R_{req}=50Ω, D_{Max}= 25%, and C_{Max}= 5%, then R_0 will be about 65Ω, while R_1 up to R_n will be in the kilo ohms range, which is much higher than R_0. This should come as no surprise since R_1 up to R_n are just used to tune the total resistance of the resistor block, while R_0 is the resistor that bears most of the current flowing in the block. Since only R_1 up to R_n have switches and not R_0, the design of the switches become really easy due to the resistors relatively high values. Consequently, the switches do not have to be designed with large sizes in order to make their resistance relatively small (switches usually need to have a relatively small resistance in order not to affect the linearity of the resistor connected to it). Moreover, not only can the switches be easily designed to have a small resistance relative to R_1 up to R_n, their contribution to the actual total resistance of the resistor block is even less since R_0 is the main contributor to the total resistance rather than R_1 up to R_n. This relaxes the design of the switches even further and reduces their noise contribution without compromising the linearity of the resistor block. Not having to design the switches to be large will also reduce the capacitive loading of the switches, which will widen the frequency range the resistor block could be used for. This advantage becomes very clear taking into account that some techniques use a single resistor instead of a resistor block, and to achieve the tuning, a bank of resistors is used with each resistor having a slightly drifted value from R_{req}. Depending on process variations, only one resistor out of the bank of resistors is used. This means that the switch has to be negligible relative to values close to R_{req}, which requires significantly larger transistors as opposed to the described technique. Note that if the switch does not have a relatively negligible resistance, it will degrade the linearity of the total resistance and will also increase the variation in the total resistance with temperature. This clarifies the advantage of the described technique and also justifies the second assumption made during the development of the recursive Eqs. 5-8 to 5-10.

The fourth advantage of the described technique is that it allows the tuning of several, different-valued resistors independently using the same calibration circuit whether those resistors are implemented in a close proximity, or scattered across the chip. If the different-valued resistors are laid out carefully and in close proximity, then simply a single R_{int} could be used to tune all those different resistors simultaneously. This is possible since resistors laid out in close proximity usually maintain their ratios to within 0.1%. If the resistor blocks have to be scattered around different locations on the chip, then there are two options. The first option is to dedicate a separate R_{int} resistor for each individual resistor block and lay it out in close proximity to its corresponding resistor block. The same control matrix and reference generator could then be used along with a state machine to perform the calibration of each resistor block successively. The state machine stores the results of the calibration of each individual resistor block and then provides independent control voltages for each block. Note that for this option, even though R_{int} is replicated, still only a single control matrix and reference generator are needed, which significantly saves area.

The second option for calibrating scattered, different-valued resistors is to use a single R_{int} resistor for all the resistor blocks, and simply reduce C_{Max} further to account for mismatch between these scattered resistor blocks. This option is preferable since it eliminates the need to replicate R_{int}, which saves even more area on the expense of reducing C_{Max}. Note that reducing C_{Max} might imply increasing the number of resistors in each resistor block, which again increases the consumed area. Generally though, since R_{int} needs to have a relatively high value to reduce the amount of current required to produce V_{Br} in Fig. 5-3, the area saved by not having to replicate R_{int} is usually larger than the area added in the resistor blocks in order to reduce C_{Max}. Since the second option does not require the use of a state machine, i.e. the calibration is performed for all the resistor blocks simultaneously and they all use the same control signals, and does not require the replication of R_{int}, this option is the preferred option especially that designing the resistor blocks for a reduced C_{Max} is much easier from a layout perspective than replicating R_{int}. If the second option is chosen though, it is desirable to implement the single R_{int} at the center of the chip in order to give an average estimation of the errors in the resistors in the rest of the chip.

Taking into consideration that the control matrix itself whether it is used in conjunction with the first or the second options could be implemented using a single comparator on the expense of using a simple state machine (if used in conjunction with the first option), or a little more complicated state machine (if used in conjunction with the second option). This way, more area could be saved. Additionally, using a single comparator guarantees a systematic error in the calibration process (comparator offset) that could be easily corrected for, instead of random errors that could be introduced due to multiple comparators. Also, using a single comparator shifts the complexity of the control matrix to the digital domain, in which the design is easier, simply synthesized, and less sensitive to process variations.

The fifth advantage is that the technique does not require any accurate voltage references. Essentially, V_B is arbitrary and does not need to be accurate. This is due to the fact that the reference voltages are generated using the same voltage V_B that is used to generate the current flowing into R_{ext}. This is an advantage for systems that do not have a band-gap circuit readily available on-chip.

The sixth advantage of the technique is that it's using the same components that are usually used to generate accurate biasing currents for analog parts in the system without disturbing it. The structure shown in Fig. 5-3 could actually be used to generate accurate biasing currents by tapping off more current mirrors. In this case V_B could be the output of a band-gap circuit, which has to be used anyway to generate the biasing currents whether the calibration technique is used or not. In some analog parts, the biasing currents do not have to be very accurate, and instead of using band-gap voltage across an external resistor, it is used across an on-chip resistor. In that case, the structure in Fig. 5-4 could be used to generate those biasing currents as well by setting R_{rt} in Eq. 5-18 to the desired value. Whether the biasing currents are generated using the band-gap voltage across an external resistor or an on-chip resistor, the tuning architecture can be easily accommodated. Hence, power consumption of the whole architecture is effectively only the power consumed in R_{int} and the control matrix.

5.2.3 Design Considerations

There are some useful considerations that should be taken into account during the design process of the described tuning architecture. First, if the application requires certain accuracy for the implemented on-chip resistance, then tighter accuracy should be targeted during the design. This is to account for errors in the value of R_{ext} (usually within 1%), any mismatch errors in the mirroring ratio K, errors encountered in the comparison process between V_{Br} and the reference voltages, and errors in the ratio between R_{int} and the rest of the resistors in the resistor block. Usually a 2 to 3 percent margin in C_{Max} is sufficient. Extra margin could be also achieved through choosing D_{Max} (the maximum error in the value of on-chip resistors) to be a little higher than what process characterization shows.

Second, it is not desirable to have to tune for errors due to temperature change since it is time-dependent. Having to tune for temperature variations will consequently require the tuning circuit to be active at all times, which increases power consumption. Even though the described technique could be activated continuously, but it is wiser to avoid it. Therefore, under nominal process conditions and across temperature range, the designer has to make sure that each resistor in the resistor block does not vary more than $\pm C_{Max}$% from the nominal value. Using silicide-block poly resistors to implement the resistors in the resistor block easily fulfills this requirement due to their very small temperature coefficient. Having a negligible switch resistance helps a lot too in minimizing variations with temperature. If there is no need to tune for temperature variations, then the tuning circuit could be activated only one time during power-up and then deactivated for the rest of the operation time without worrying about temperature effects.

Third, if the required on-chip resistance is small (50Ω for example), then R_{int} has to be highe in order to save power and to create a high enough V_{Br} for adequate comparison with V_B. Otherwise if R_{int} is chosen to be 50Ω, the current needed to generate V_{Br} will be significantly higher, and accuracy will also suffer due to the significant mismatch error in current mirrors with large mirroring ratio. The designer has to be careful to design R_{int} and all the resistors in the resistor block to have wide dimensions and are laid out in close proximity to guarantee that all of them are affected equally (percentage wise) with process variation even though their values are different. Using poly resistors significantly improves the tuning accuracy due to their good matching characteristics (versus well resistors for example). Also using a unit resistor to implement R_{int} and the rest of the resistors in the resistor block also improves accuracy even further. However, area penalty has to be paid for using poly resistors as well as unit resistors. Therefore the designer has to reach a compromise between accuracy and area.

5.3 Applications

Integrating transmission line terminations of high-speed wire line transceivers has very attractive advantages. First, it minimizes the number of passive components on the printed circuit board, which leads to smaller boards and less cost. Second, in high-speed systems, termination resistors have to be implemented as close as possible to the chip, which significantly complicates the board layout. Integrating the termination resistors eliminates this difficulty. Third, as data rates become higher, the effect of the package parasitics becomes very significant to the extent that integrating those terminations might be the only way to achieve adequate performance. However, integrating those terminations is a challenge due the loose accuracy of on-chip resistors, which leads to high reflection coefficients especially in fast data rates[14,19]. Thus, most high-speed transceivers require a minimum accuracy for adequate performance. Many techniques exist in the literature for implementing accurate termination resistors. Some techniques use purely-active devices with an analog tuning loop to implement the required terminations[5,20]. The implemented resistors usually suffer from linearity, range, and noise limitations, in addition to the disadvantages of using analog tuning as discussed earlier. Other techniques use purely-active devices along with a digital tuning loop to implement the required terminations[21]. Those techniques, even though have the advantage of digital tuning, yet, the terminations are still implemented using purely-active devices, therefore, linearity, range, and noise limitations still exist. Other techniques use a mixed combination of active elements and passive on-chip resistors along with an analog tuning loop[4,14]. Even though, those techniques achieve better linearity than the purely-active approach, yet they still suffer from range and noise limitations since active elements still constitutes a significant portion of the implemented resistance, in addition to the disadvantages of analog tuning. On the other hand, the described technique in this chapter has the advantages of digital tuning, better linearity (almost as good as a passive on-chip resistor), better noise performance, and wider operation range. In order to demonstrate the use of the described tuning architecture in general, and in the area of integrated transmission line terminations in particular, the following sections show the implementation of on-chip transmission line terminations for two high-speed wire line digital transceivers. The first transceiver operates at 480 Mbps and has only one port. Thus, it requires one pair of terminations (2 resistor blocks). The second transceiver operates at 1.65Gbps and has 8 ports. Thus, it requires 8 pairs of terminations (16 resistor blocks).

5.3.1 A 480 Mbps Transceiver

The transceiver specifications require two single-ended termination resistors. Each resistor is 45Ω with a tolerance of no more than ±10%, and the targeted implementation process for the transceiver is a standard 0.18μm technology. Following the procedure described earlier in section 5-1 for implementing accurate on-chip resistors, C_{Max} is chosen to be 8.67% in order to give a 1.33% margin to account for any possible mismatch errors. Process characterization shows that the maximum change from the nominal value for on-chip silicide-block poly resistors is ±25% from the nominal value including variations due to temperature changes as well, which implies that D_{Max} should be 25%, but for extra margin, D_{Max} is chosen to be 26%. Using Eq. 5-1, the number of parallel resistors needed to achieve this accuracy is found to be $n = 4$, therefore, and by using Eqs. 5-8 to 5-10, the range of values of each resistor in the resistor block is calculated, and is shown in Table 5-1. As discussed before, since the average value of the range allowed for each resistor gives the maximum margin, the average value is simply used as the nominal value for each resistor. Note how the modulating resistors R_1 through R_4 are order of magnitude higher than the required total resistance (45Ω). Since the values in Table 5-1 show the total resistance, the resistance of the switches has to be accounted for. Hence, the total resistance is divided between the switch and an actual silicide-block poly resistor, where the switch resistance is designed to be around 10% of the total resistance. Note that the contribution of the switches to the total resistance of the resistor block will be even much less (less than 2%). Table 5-2 shows the resistance of each silicide-block poly resistor along with the switch resistance. In the implementation of this transceiver, accurate biasing currents are needed anyway, hence, a band-gap circuit and an off-chip resistor R_{ext} are readily available. Therefore, the output of the band-gap circuit is used V_B, while R_{ext} is chosen to be 6.3 KΩ. The mirroring factor K is simply chosen to be unity, which makes the nominal value of R_{int} equals to 6.3 KΩ as well. Note that the supply used for the band-gap circuit as well as the current mirrors is 3.3V±10%. Each resistor in the resistor block as well as R_{int} is checked under nominal conditions and temperature range from -40 to 125 ^{O}C to make sure that the relative magnitude change in the value of each resistor with temperature variations is less than ± 8.67%. This is easily

Table 5-1. Range of values the resistors in the resistor block can take for implementing 45Ω±8.67% terminations for the 480Mbps transceiver.

Resistor	Minimum (Ω)	Maximum (Ω)	Average (Ω)
R_0	56	59	57.5
R_1	374	806	590
R_2	323	323	323
R_3	238	564	401
R_4	262	953	607.5

Table 5-2. Values of the resistors in the resistor block divided between switches and silicide-block poly resistors for the 480Mbps transceiver.

Resistor	Silicide-block resistance (Ω)	Switch resistance (Ω)	Total resistance (Ω)
R_0	57.5	N/A	57.5
R_1	531	59	590
R_2	290.7	32.3	323
R_3	360.9	40.1	401
R_4	546.75	60.75	607.5

achievable due to the low temperature coefficient of silicide-block poly resistors. The reference generator resistors are calculated using Eq. 5-18 with R_{rt} chosen to be 6.3 KΩ. R_{r0} is found to be 5.208 KΩ, while the rest of the resistors are found to be all equal to 546 Ω. Since for this application the tuning time is not critical, the control matrix is implemented using a single comparator and a state machine rather than the flash architecture shown in Fig. 5-5. Also the current mirror in Fig. 5-3 is implemented using a low-voltage cascode. Since there is only two identical resistor blocks implemented, they are laid out in the same orientation and in close proximity to each other and to R_{int} as well, in order to minimize any mismatch errors, while the switches are implemented using NMOS transistors. After laying out the two resistor bocks, it is found that each resistor block occupied a total area of 206 μm^2, while R_{int} occupied an area of 176 μm^2. The reference generator including the resistors as well as the Op-amp and the source follower transistor is found to occupy an area of 3229 μm^2, while the comparator occupied an area of 6545 μm^2. The total area of the complete design including the terminations, the reference generator, and the control matrix represented about 3.6% of the complete transceiver. To verify the design, simulations across different process corners, temperature variations (-40 to 125 ^{o}C), and different supply voltages (3.3V ±10%) are performed. Additionally, statistical simulations are performed as well to take into account errors due to mismatches between resistors, current mirrors, offset introduced by op-amps and comparators in the control matrix, and process variations. All the previous simulations are done taking into account a ±1% error in R_{ext}. Simulation results show that the total resistance of the resistor block is well controlled within 41.25 Ω and 48.82 Ω, with a nominal value of 44.8 Ω, which is about ±8.5% from 45Ω. The total power the calibration procedure consumes is around 1.27 mW, and it is performed only during the startup of the transceiver, and repeated every time the transceiver goes through a reset protocol, otherwise it is disabled. To further verify the design on silicon, the whole transceiver including termination resistors was implemented on a standard 0.18μm technology and measurements were performed on the terminations for different process corners. The resistance was measured at different voltage levels up to 70% of the 3.3V supply voltage (≈2.3V), and the maximum error from the resistance value measured at the middle of this range (35% of the 3.3V supply ≈1.1V) was found to be within ±0.017%, which demonstrate the highly linear performance of the resistor block over a wide voltage range as expected by theory in section 5-1. Table 5-3 shows measurement results of the implemented resistors, which align very well with simulation results, while Fig. 5-7 shows the value of the resistance as a function of the normalized voltage applied across it (for perfectly linear resistors, the resistance would be constant). As shown in figure, the linearity error is very small and the value of the resistance is within design expectations.

Table 5-3. Measured resistance implemented using the proposed architecture.

	-3σ	Mean	+3σ	Error from 45Ω
Fast Process	42.17Ω	45.41Ω	48.65Ω	-6.3% to +8.1%
Nominal Process	42.71Ω	45.77Ω	48.83Ω	-5.1% to +8.5%
Slow Process	43.11Ω	45.81Ω	48.51Ω	-4.2% to +7.8%

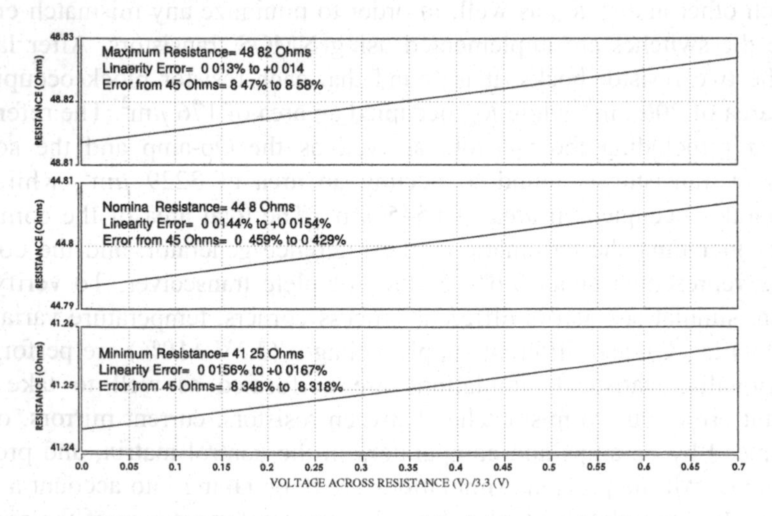

Figure 5-7. Linear performance of the implemented resistance.

5.3.2 A 1.65 Gbps Transceiver

Specifications for this transceiver require termination resistors to be 50 Ω with a tolerance of no more than ±10%. The targeted process for the transceiver is a standard 0.13μm technology. For this transceiver though, and since it has 8 different ports, the required 16 termination resistors could not be implemented in close proximity to each other. In fact, they have to be scattered across different locations on the chip. Therefore, and as discussed before, there are two approaches to implement the tuning architecture. The first approach is to tune each pair of termination resistors individually, which requires that each termination resistors pair has to have its own R_{int} as well as its own control matrix, and consequently consume more area and power. The second approach is to reduce C_{Max}, i.e. tune to a tighter accuracy in order to provide margin to account for the scattered locations of the resistors, and use the same R_{int} and control matrix for all 8 termination resistors pairs. In order to achieve that, C_{Max} is chosen to be 5% instead of 10%. Process characterization shows that the maximum change from the nominal value for on-chip silicide-block poly resistors is ±25% including variations due to temperature, thus D_{Max} is chosen to be 25%. Note that D_{Max} is not chosen to be higher than 25 since the reduction in C_{Max} (5%) is enough margin. Again using Eq. 5-1, the number of parallel resistors in each resistor block needed to achieve this accuracy is found to be $n = 8$, and by using Eqs. 5-8 to 5-10, the range of values each resistor in the resistor block can take is calculated and is shown in Table 5-4. The average value of the range allowed for each resistor gives the maximum margin, and is simply used as the nominal value for each resistor (note the higher value of the modulating resistors). The division between the silicide-block resistors and the switches is again chosen to be 10% as shown in Table 5-5. As in the 480 Mbps transceiver, a band-gap circuit and off-chip resistor R_{ext} are readily available, and are thus used.

Table 5-4. Range of values the resistors in the resistor block can take for implementing 50Ω±5% terminations for the 1.65Gbps transceiver.

Resistor	Minimum (Ω)	Maximum (Ω)	Average (Ω)
R_0	63.33	65.63	64.48
R_1	750.06	1467.16	1108.61
R_2	673.94	1366.23	1020.08
R_3	642.62	1420.44	1031.53
R_4	610.00	610.00	610.00
R_5	475.00	1050.00	762.50
R_6	541.59	1685.53	1113.56
R_7	499.74	1544.06	1021.90
R_8	481.91	1609.12	1045.52

Table 5-5. Values of the resistors in the resistor block divided between switches and silicide-block poly resistors for the 1.65Gbps transceiver.

Resistor	Silicide-block resistance (Ω)	Switch resistance (Ω)	Total resistance (Ω)
R_0	65	N/A	65
R_1	997.2	110.8	1108
R_2	918	102	1020
R_3	928.8	103.2	1032
R_4	549	61	610
R_5	686.7	76.3	763
R_6	1002.6	111.4	1114
R_7	919.8	102.2	1022
R_8	941.4	104.6	1046

The same values of R_{ext}, R_{int}, K, and R_{rt} used in the 480 Mbps transceiver are also used for this transceiver. Using Eq. 5-18 with R_{rt} chosen to be 6.3 KΩ, R_{r0} is found to be 5.04 KΩ, while the rest of the resistors are found to be all equal to 315 Ω. To verify the design, simulations are performed with -40 to 125 ^{O}C temperature range, different supply voltages (3.3V ±10%), and ±1% error in R_{ext}, along with statistical simulations. Results show that the total resistance of all resistor blocks (total of 16) is well controlled within 46.76 Ω and 53.26 Ω, which is about ±6.5% from 50 Ω. Note that the ±6.5% includes only short-range mismatches between the resistor blocks. Taking into account that the resistor blocks are scattered across the chip, 2% extra loss in accuracy is roughly estimated.

5.4 Summary

This chapter presented a digital adaptive technique that maintains a tight control over on-chip resistors. The technique uses a single off-chip resistor as a reference and does not require an accurate voltage reference. It could be used as a master calibrator for several, different-valued on-chip resistors even if they are scattered around the chip. The accuracy of the technique is limited by the accuracy of the reference resistor and mismatch errors. The architecture maintains low noise, high linearity, wide voltage range, and high-speed performance of the implemented resistance. In addition to the concept, a mathematical methodology that makes the whole technique simple, fast, and practical was also presented. The technique could be used to design accurate on-chip resistors for different applications such as on-chip terminations, filters, oscillators, and most applications that require accurate resistors. Terminations for two high-speed wire line transceivers (480 Mbps, and 1.65 Gbps) that require ±10% accuracy were implemented using this technique. Simulation and measurement results were presented and showed adequate performance across process, temperature, and supply ranges.

LIST OF REFERENCES

1. Mohammed Ismail, Terri Fiez, "Analog VLSI Signal and Information Processing," McGraw-Hill, New York, 1994.
2. T. J. Gabara, S. C. Knauer, "Digitally adjustable resistors in CMOS for high-performance applications," *IEEE J. Solid-State Circuits*, vol. 27, pp. 1176-1185, Aug. 1992.
3. A. DeHon, T. Knight, Jr., T. Simon, "Automatic impedance control," *IEEE International Solid-State Circuits Conference*. Digest of Technical Papers, pp. 164-165, 283, Feb. 1993.
4. Thaddeus Gabara, Wilhelm Fischer, Wayne Werner, Stefan Siegel, Makeshwar Kothandaraman, Peter Metz, and Dave Gradl, "LVDS I/O Buffers with a Controlled Reference Circuit," *Proceedings of the 10th Annual IEEE International ASIC conference*, pp. 311-315, Sept. 1997.
5. Hongjiang Song, "Dual mode transmitter with adaptively controlled slew rate and impedance supporting wide range data rates," *Proceedings of the 14th Annual IEEE International ASIC/SOC conference*, pp. 321-324, Sept. 2001.
6. M. Ismail, S.V. Smith, and R.G. Beale, "A new MOSFET-C universal filter structure for VLSI," *IEEE J. Solid-State Circuits*, vol. SC-23, pp. 183-194, Feb. 1988.
7. S. Sakurai, and M. Ismail, "A CMOS square-law programmable floating resistor independent of the threshold voltage," *IEEE Trans. Circuits Syst. II*, vol. 39, pp. 565-574, Aug. 1992.
8. R. Schaumann, M. S. Ghausi, and K. R. Laker, "Design of Analog filters, passive, active RC, and switched-capacitor", Prentice-Hall, Englewood Cliffs, NJ, 1990.
9. Satoshi Sakurai, Mohammed Ismail, Jean-Y ves Michael, Edgar Sanchez-Sinencio, Robert Brannen "A MOSFET-C Variable Equalizer Circuit with Simple On-Chip Automatic Tuning," *IEEE Journal of Solid-State Circuits*, vol. 27, NO. 6, June 1992.
10. K. Nagaraj, "New CMOS floating voltage-controlled resistor," *Electronics Letters*, vol. 22, PP. 667-668, 1986.
11. S. P. Singh, J. V. Hanson, and J. Vlach, "A new floating resistor for CMOS technology," *IEEE Trans. Circuits. Syst.*, vol. 36, PP. 1217-1220, Sept. 1989.
12. M. Steyaert, J. Silva-Martinez, and W. Sansen, "High-frequency saturated CMOS floating resistor for fully-differential analog signal processors," *Electronics Letters*, vol. 27, PP. 1609-1611, 1991.
13. Behzad Razavi, "Design of Analog CMOS Integrated Circuits," McGraw-Hill, New York, 2001.
14. H. Conrad, "2.4 Gbit/s CML I/Os with integrated line termination resistors realized in 0.5/spl mu/m BiCMOS technology," *Proceedings of the Bipolar/BiCMOS Circuits and Technology Meeting*, pp. 120-122, Sept. 1997.
15. David A. Johns, Ken Martin, "Analog Integrated Circuit Design," John Wiley & Sons, New York, 1997.
16. Ayman A. Fayed, and M. Ismail "A Digital Tuning Algorithm For On-Chip Resistors," *Proceedings of the 2004 International Symposium on Circuits and Systems,* Vol. 1, pp. 936-939, May 2004.
17. Ayman A. Fayed, and M. Ismail "A Digital Calibration Algorithm for implementing accurate On-Chip Resistors," *Int. J. of Analog Integrated Circuits and Signal Processing*, Accepted, Nov. 2005.
18. Frode Larsen, Mohammed Ismail, and Christopher Abel, "A versatile structure for On-Chip extraction of resistance matching properties," *IEEE Transactions on Semiconductor Manufacturing*, vol.9, No. 2, PP. 281-285, Feb. 1988.

19. I. Novak, “Modeling, simulation, and measurement considerations of high-speed digital buses,” *Instrumentation and measurement Technology Conference,* pp. 1068-1074, May. 1992.
20. T.J. Gabara, “On-chip terminating resistors for high-speed ECL-CMOS interfaces,” *Proceedings of the Fifth Annual IEEE International ASIC Conference and Exhibit,* pp. 292-295, Sept. 1992.
21. Kyoung-Hoi Koo, Jin-Ho Seo, Myeong-Lyong Ko, and Jae-Whui Kim, “Digitally tunable on-chip resistor in CMOS for high-speed data transmission,” *ISCAS 2003. Proceedings of the 2003 International Symposium on Circuits and Systems*, vol. 1, pp. 185-188, May 25-28, 2003.

Chapter 6
EQUALIZATION

The main topic of the previous chapters was variations on the circuit level and the available adaptive techniques to combat those variations. As discussed in chapter 1, circuit level variations are variations encountered during the design of each individual circuit block in a system. Those variations are essentially the variations in the absolute values of the basic parameters of devices (resistors, capacitors, and transistors) due to process, supply voltage, and temperature variations. This chapter focuses on variations on the network level, i.e. variations in the transmission media. As mentioned in chapter 1, one of the major problems with any transmission media used for digital communications is Intersymbol Interference, ISI for short. This chapter will give an overview of the causes of ISI and its effect on the reception quality of digital data. The available techniques for combating ISI, particularly equalization, including fixed and adaptive equalization will be discussed. A comparison between analog and digital implementations of equalization techniques will be presented along with the advantages and disadvantages of both methodologies. Chapter 7 will present the implementation of an analog adaptive equalizer for wire line transceivers.

6.1 Intersymbol Interference

ISI has been traditionally a serious limitation on data rates that can be sent across a specific communication channel[1-4]. ISI generally refers to the interference that occurs between the current received bit and other previously received bits in the same data stream due to the addition of portions of the previously received bits to the bit that is being received. In the simplest form, front-end digital receivers are just comparators that compare the received amplitude to a certain threshold to decide if the received bit is a 1 or a 0, therefore if the portions added from the previous bits to the current bit are enough to change the current bit from a 1 to a 0 or vice versa, then ISI could significantly affect the bit error rate of the whole receiver. On the other hand, if the portions added from the previous bits to the current bit are not enough to mislead the receiver, then theoretically it should be tolerable. However, this is not true all the time due to design limitations of the front-end receiver itself as will be discussed later. In fact, ISI could still be a potential problem even if it is not enough to change the current received bit. In order to understand the nature of ISI, its causes have to be clearly explained. As mentioned before, there has to be a way for portions of previously received bits to contaminate the current received bit. In wireless networks, the multi-path effect shown in Fig 6-1 is a straightforward cause of ISI. Basically, multiple indirect delayed versions of the transmitted signal (that usually results from multiple reflections) interferes with the direct signal received by the antenna causing the current received symbol to be contaminated with previous symbols, which is essentially ISI. Luckily though, since those delayed versions of the transmitted signal are a result of reflections, they travel for a longer distance than the direct received signal before they get to the receiver's antenna. Therefore, they are usually attenuated. If the current received symbol is X_{R_i}, then it can be represented by:

$$X_{R_i} = \alpha_i X_{T_i} + \alpha_{i-1} X_{T_{i-1}} + \alpha_{i-2} X_{T_{i-2}} + \dots\dots\dots \qquad (6\text{-}1)$$

where X_{Ti} is the transmitted symbol, and α_i is the associated attenuation factor due to the channel path. Equation 6-1 can be rewritten more generally using Z domain as:

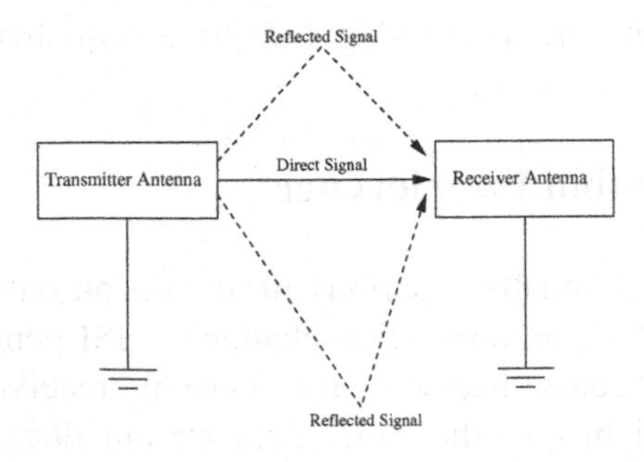

Figure 6-1. The multi-path effect in wireless networks.

$$X_R(Z) = \left(\alpha + \alpha_1 Z^{-1} + \alpha_2 Z^{-2} + \alpha_3 Z^{-3} + ...\right) X_T(Z) \quad (6\text{-}2)$$

where $X_R(Z)$ and $X_T(Z)$ are the received and transmitted signals respectively. An important thing to notice from Eq. 6-2 is that the transfer function between the transmitted and the received signals is a low pass function[3].

In wire line communications, there is really no multi-path effect since there is only one path for the signal between the transmitter and the receiver, yet, ISI could also happen due to a completely different phenomenon, which is dispersion. In order to elaborate on this, let's assume that the wire line channel could be modeled as a simple RC network, which is essentially a low-pass filter. Furthermore, let's assume that the channel has a much smaller bandwidth than the digital signal that is being sent across it and that the digital pulses are simple gate pulses. That is, a 1 is represented by a pulse with amplitude α and duration T_b, while a 0 is represented by a pulse with amplitude $-\alpha$ and the same duration. Since an RC filter is a linear filter, its response for a train of pulses could be studied by adding up its response to each individual pulse. Figure 6-2 shows the response of a band-limited channel to a single pulse. As shown, the response of the channel to such a pulse involves two effects. First, the maximum amplitude the output pulse can reach is less than the original amplitude of the input pulse. Second, the output pulse becomes dispersed and its duration extends beyond T_b. The rising edge of the output pulse could be written as:

$$V_O(t) = a\left(1 - e^{-t/RC}\right) \quad \text{for} \quad t < T_b \quad (6\text{-}3)$$

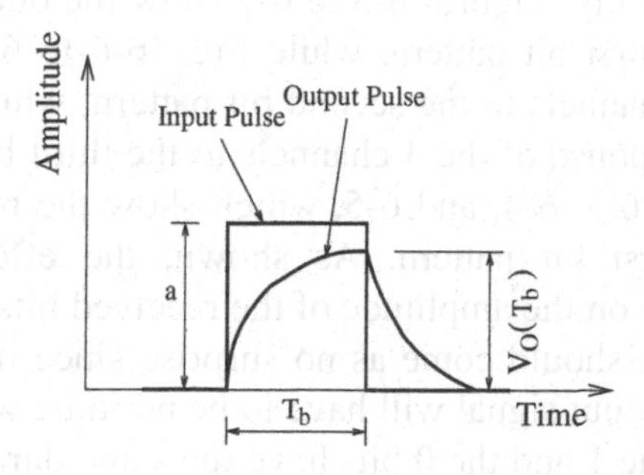

Figure 6-2. The response of a band-limited channel to a single pulse.

Using Eq. 6-3, the maximum voltage in the output pulse could be found by replacing t with T_b, that is:

$$V_O(T_b) = a\left(1 - e^{-T_b/RC}\right) \tag{6-4}$$

On the other hand, using Eq. 6-4, the falling edge of the output pulse could be written as:

$$V_O(t) = a\left(1 - e^{-T_b/RC}\right)e^{-t/RC} \quad \text{for} \quad t > T_b \tag{6-5}$$

As Eqs. 6-3 to 6-5 show, the output pulse is being spread out in time beyond its allocated time period T_b, therefore, when another pulse is sent out across the channel, the residue from the previous pulse represented by Eq. 6-5 is going to contaminate the new received pulse, and potentially mislead the front-end receiver[5].

The severity of ISI introduced by the low-pass nature of the communication channel is a function of two factors. The first factor is the bandwidth of the channel, while the second factor is the density of transitions in the data stream being sent across the channel, i.e. the bit pattern. Therefore, ISI is usually categorized as a deterministic jitter since it could be related back to the bit pattern. In order to clarify the effect of both factors, the following experiment could simply be performed. Let's assume that the data rate of the transmitted data is 500 Mbps, i.e. bit duration is 2 ns. Let's also assume that a 1 is a gate pulse with amplitude 1 V, while a 0 is a gate pulse of amplitude –1 V. Furthermore, let's assume we have 3 different channels, channel 1 has a bandwidth of 250 MHz, channel 2 has a bandwidth of 125 MHz, and channel 3 has a bandwidth of 62.5 MHz. Let's also assume the following three bit patterns: The first pattern is an alternating 1-0 pattern, i.e. a 1 followed by a 0 that repeats periodically, the second pattern is 1-1-1-0 that also repeats periodically, and the third pattern is 1-1-1-1-1-1-1-0 that also repeats periodically. Figures 6-3 to 6-5 show the output response of the 3 channels for the first bit pattern, while Figs. 6-6 to 6-8 show the output response of the 3 channels to the second bit pattern, while Figs. 6-9 to 6-11 show the output response of the 3 channels to the third bit pattern. First off, let's compare Figs. 6-3, 6-4, and 6-5, which show the response of all three channels to the first bit pattern. As shown, the effect of the channel bandwidth is mainly on the amplitude of the received bits, while the duration is not affected. This should come as no surprise since if the input signal is periodic, then the output signal will have to be periodic with the same period as well, and since the 1 and the 0 bits have the same duration, then they will appear as such at the output too. Therefore, for an alternating 1-0 pattern, the effect of ISI appears only as attenuation in the amplitude of the received bits, but not on their time duration. This is an important result since it can be used to differentiate between ISI and different kinds of jitter at the output of a specific channel. Specifically, it is a common practice to send an alternating 1-0 pattern across a communication channel in order to exclude ISI from jitter measurements. Second, by comparing Figs. 6-4, 6-7, and 6-10, which show the response of channel 2 to all three different patterns, it is obvious that the 0 bit has a different amplitude and duration even though it is going through the same channel. It is also obvious that the longer the identical series of 1s before the 0 is, the worse the effect of ISI on the amplitude and duration of the 0 bit is. This also should come as no surprise since the longer the input stays at constant amplitude, the more the channel charges to that amplitude, which will make it more difficult to discharge when the input switches. Therefore, ISI starts to affect the duration of the received bits only if the input pattern is not symmetrically alternating as in the first bit pattern. Third, by comparing Figs. 6-6, 6-7, and 6-8, as well as 6-9, 6-10, and 6-11, which show the response of the 3 channels to both the second and third bit patterns, it is obvious that the lower the channel bandwidth is, the worse the effect of ISI on the received bits amplitude and duration, given that the input bit pattern is not a symmetric alternating pattern. This is also expected since the slower the channel is, the more dispersed the output pulses get, and consequently the more they interfere with each other. In fact, as shown in Fig. 6-11, the effect of ISI on the output pattern is so severe that the 0 bit does not even reach the threshold of the front-end receiver, which will essentially be interpreted as a 1 instead of a 0.

To summarize the previous discussion, there are two conclusions. The first conclusion is that as the channel bandwidth gets lower, the effect of ISI becomes higher on both the amplitude and duration of the received bits, with the exception of symmetric alternating patterns, where the effect of ISI appears only on the amplitude of the received bits but not their duration. The second conclusion is that as the series of identical bits before a complementary bit is sent gets longer, the effect of ISI on the amplitude and duration of the received complementary bit becomes higher. One way to fight that is bit stuffing, which is used in NRZ transmission schemes, where a complementary stuffing bit is inserted when the data stream includes a series of identical 1s or 0s that are longer than a specific length. Bit stuffing is very important in facilitating the clock recovery from NRZ data streams, as well as to minimize the effect of ISI.

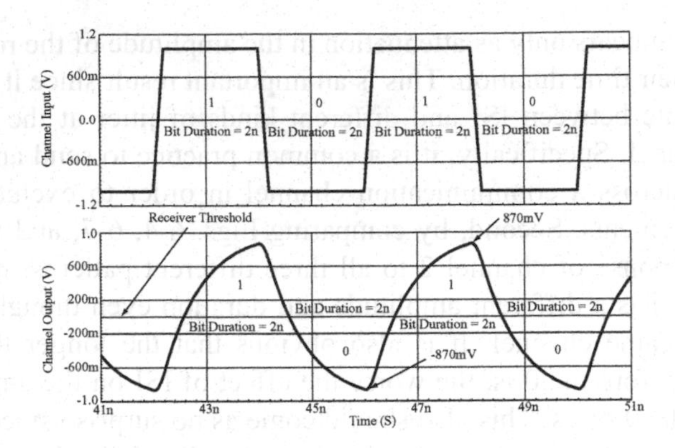

Figure 6-3. Response of channel 1 to the first input pattern.

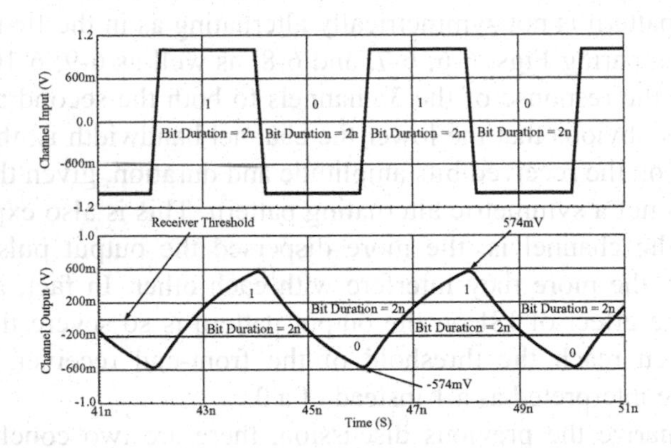

Figure 6-4. Response of channel 2 to the first input pattern.

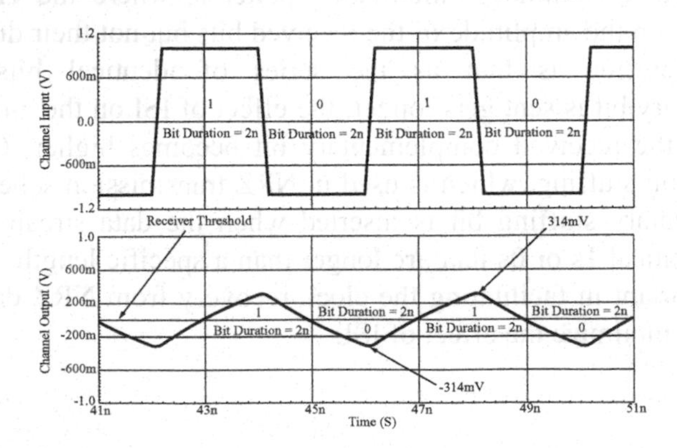

Figure 6-5. Response of channel 3 to the first input pattern.

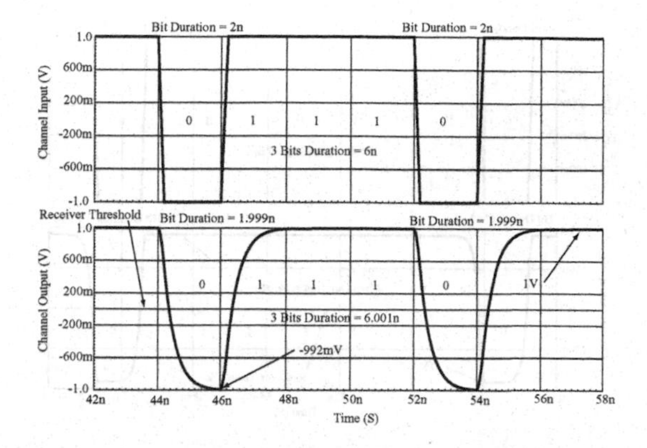

Figure 6-6. Response of channel 1 to the second input pattern.

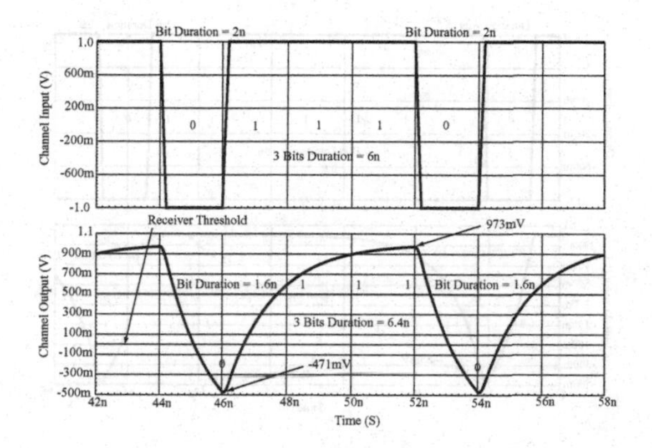

Figure 6-7. Response of channel 2 to the second input pattern.

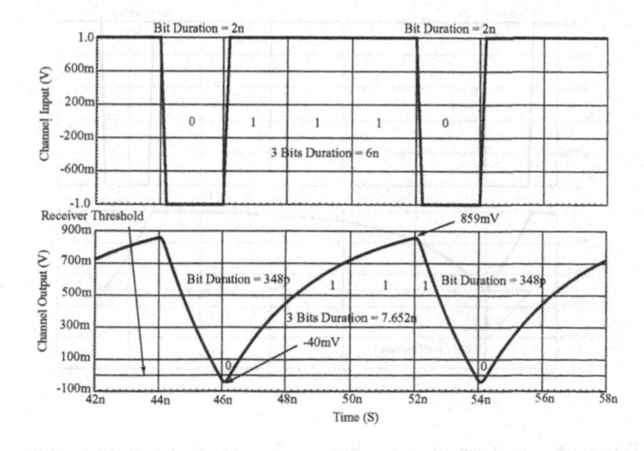

Figure 6-8. Response of channel 3 to the second input pattern.

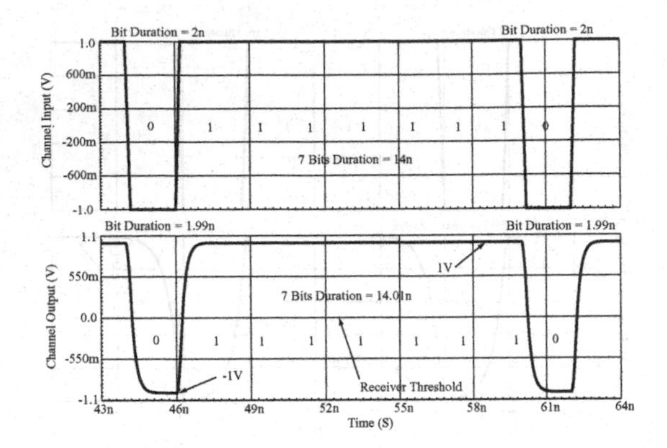

Figure 6-9. Response of channel 1 to the third input pattern.

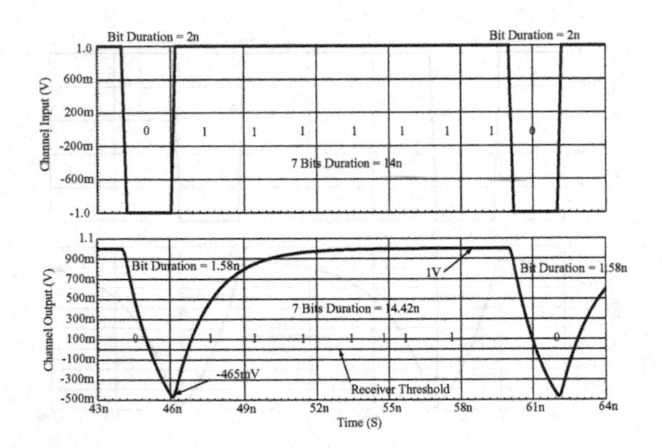

Figure 6-10. Response of channel 2 to the third input pattern.

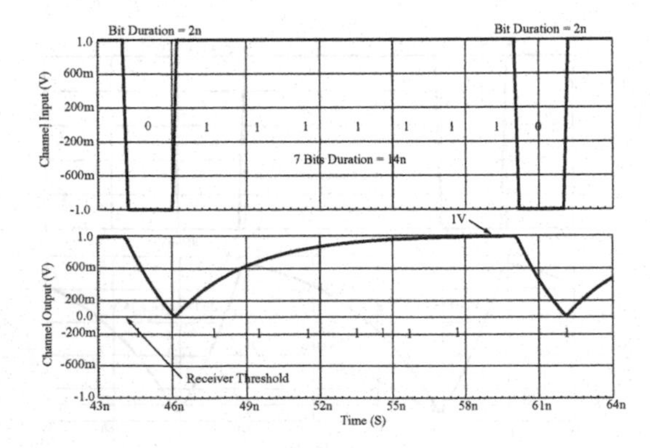

Figure 6-11. Response of channel 3 to the third input pattern.

6.2 Eye Diagrams

As discussed in the previous section, ISI is a function of the bit pattern being sent across the channel. If the input pattern is very long and random, it becomes a very difficult task to find out the effect of ISI on both the amplitude and duration of the received bits. A good tool to assess ISI (and jitter in general) in the received data stream is eye diagrams[3]. Eye diagrams are created by sending a continuous random data stream and then capturing the output wave form. The captured waveform is then divided to sections with duration equal to the ideal bit period, and then those sections are overlaid on top of each other to create an eye diagram. The two important parameters in an eye diagram are the vertical and horizontal openings of the eye. The vertical opening represents the minimum amplitude the received bits can have. It usually defines the sensitivity of the front-end receiver, i.e. the minimum amplitude the receiver has to be able to detect, which by definition is the minimum gain of the receiver. The horizontal opening on the other hand represents the minimum duration the received bits could have. Hence, it usually defines the speed of the receiver, i.e. the minimum bit width the receiver has to be able to detect. In order to illustrate the horizontal and vertical openings of an eye, Figs. 6-12 to 6-14 show the three eye diagrams that result from sending a long random 500Mbps data stream across channels 1, 2, and 3 defined earlier in the previous section. Ideally, the horizontal eye opening of such data stream should be 2ns, while the vertical opening should be 2V. As shown in figure, the first eye diagram that represents the response of channel 1 has the widest vertical and horizontal openings since it has the highest bandwidth. As the bandwidth gets lower, both the vertical and horizontal eye openings get smaller. At the extreme case of channel 3 shown in Fig. 6-14, the eye gets completely closed, which essentially means that ISI was severe enough that some bits could not even cross the threshold of the front-end receiver and therefore will be received incorrectly.

In theory, as long as there is a vertical and horizontal eye opening, then it is possible to detect the received bits correctly since ISI was not severe enough to prevent the bits from reaching the threshold of the front-end receiver. If the eye is completely closed as shown in Fig. 6-14, then obviously the received data stream can not be received correctly without further processing. Unfortunately though, this is only true to a limit. The reason behind that is that ISI affects both the vertical and horizontal openings simultaneously, which essentially means that when ISI becomes more severe, the front-end receiver has to achieve higher gain and higher speed at the same time. Since it is a well known fact that the gain and bandwidth of a comparator are two contradictory parameters, i.e. for the same power consumption, higher gain means lower bandwidth and vice versa[6-9], then for very small eye openings, even though there is an opening, it becomes very difficult to design a front-end receiver with acceptable power consumption. At high data rates, this problem becomes significant.

Whether the eye is completely closed or the vertical and horizontal openings are too narrow, further processing is needed to widen the eye opening in the received signal, i.e. compensate or eliminate the effect of ISI. The technique used to combat the effect of ISI is referred to as equalization since it equalizes the effect of the transmission channel to enable the receive path to achieve higher bit error rates. In the following section an overview of equalization techniques and architectures will be presented.

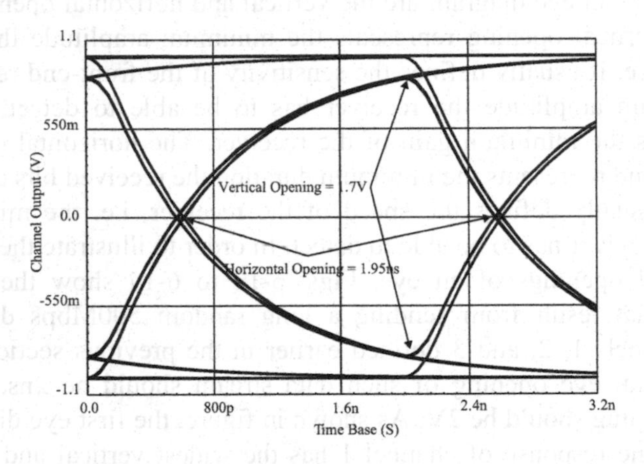

Figure 6-12. The Eye diagram of a random data stream across channel 1.

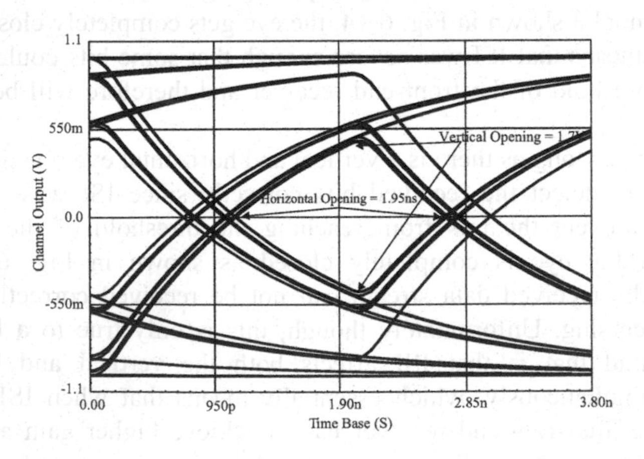

Figure 6-13. The Eye diagram of a random data stream across channel 2.

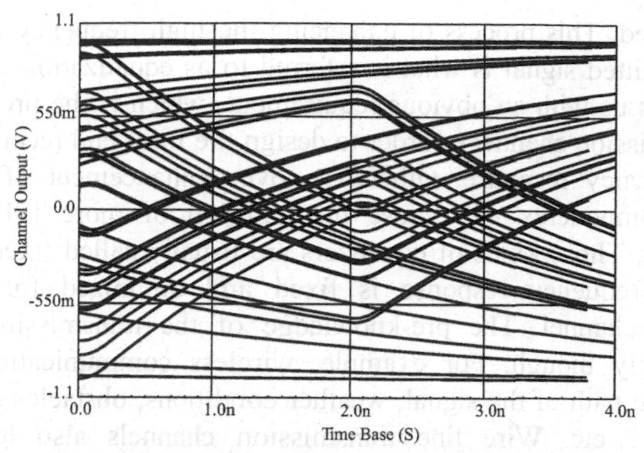

Figure 6-14. The Eye diagram of a random data stream across channel 3.

6.3 Equalization Architectures

Whether ISI is introduced through multi-path effects or through dispersion due to the limited bandwidth of the transmission media, the transfer function between the transmitted and received data streams could be modeled as a bandwidth limited low-pass transfer function[3]. In other words, for a given data stream, ISI results from the unequal attenuation and delay the different frequency components of the transmitted signal suffer from due to the low-pass nature of the transmission channel. In fact, the relatively higher attenuation that the high frequency components of the transmitted signal suffer from when compared to the lower frequency components is the main reason for the degradation in the horizontal eye opening of the received signal since those higher frequency components are the ones that represent the fast rising and falling edges of the transmitted signal. Therefore, when they get attenuated, the received signal losses those fast rising and falling edges. Given the explanation above, it becomes obvious that the straight forward cure for ISI is the design of the front-end receiver such that it relatively enhances the high frequency components of the received signal in order to compensate for the higher attenuation they suffered from due to the low-pass transmission channel. In other words, the front-end receiver needs to have a high-pass transfer function with a unity DC gain to compensate for the low-pass nature of the channel[1]. Essentially, what this enhancement does is that it increases the effective bandwidth of the transmission channel. In order to illustrate the idea, Fig. 6-15 shows the transfer function of a low-pass transmission channel, the transfer function of a high-pass front-end receiver, and their combined frequency response. As shown in figure, the effective bandwidth of the transmission channel with the high-pass filter is almost doubled. This process of enhancing the high frequency components of the transmitted signal is what is referred to as equalization. Figure 6-15 though leaves us with an obvious requirement, which is the pre-knowledge of the transmission channel in order to design the front-end receiver with the correct frequency response. Otherwise, over enhancement of the higher frequency components will cause the addition of more ISI instead of eliminating it. Those kinds of equalizers are usually called fixed equalizers since their frequency response is fixed and optimized for a specific transmission channel. The pre-knowledge of the transmission media is highly unlikely though. For example, wireless communication channels depend on the path of the signal, weather conditions, obstacles or buildings in the way ...etc. Wire line transmission channels also have similar problems, i.e. Cable length, connectors' types, temperature ...etc. This makes it very difficult to predict the frequency response of the transmission media, and therefore using a fixed equalizer might not be adequate. For the above reasons, adaptive equalizers are widely employed, in which the frequency response of the front-end receiver is automatically adapted to the transmission channel. There are many different equalization architectures that could be implemented in the analog domain (continuous or discrete-time), or in the digital domain. In the next few sections, an overview of the different equalization architectures will be presented.

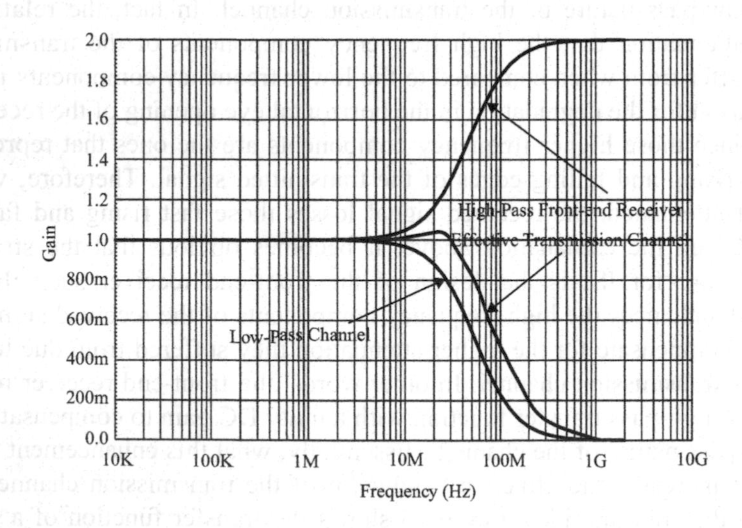

Figure 6-15. The effect of enhancing the high frequency components of the received signal on the effective bandwidth of the transmission channel.

6.3.1 The Feed-Forward Architecture

The first equalization architecture is the feed-forward architecture shown in Fig. 6.16. This architecture is simply based on the fact that if the total transfer function between the transmitter and the receiver has enough bandwidth, then the transmitted waveform will retain its shape and go through without any ISI[3]. As mentioned earlier, in order to achieve that, the front-end receiver has to have some sort of a high-pass response with a unity DC gain to compensate for the low-pass response of the transmission channel, hence restoring the transmitted signal shape and pass it to the digital receiver (the slicer). Therefore, the feed-forward architecture shown in Fig. 6-16 uses one high-pass filter as its front-end receiver, which is also referred to as the Forward Equalizer (FE). An important aspect of this feed-forward architecture is that it only tries to remove the effect of ISI by restoring the original shape of the received waveform through enhancing the speed of transitions (high-pass response). Therefore, it does not use any knowledge of the previously received bits in the data stream to remove ISI, i.e. there is no feedback. This is the main reason for calling forward equalizers ISI pre-cursor removers[10].

The feed-forward architecture has the major advantage of restoring the shape of the transmitted waveform, or in other words, it widens up the horizontal eye opening of the received signal. As discussed in the previous section, this essentially makes the design of the slicer much easier. The other advantage of the feed-forward architecture is its open loop nature, i.e. it does not require the knowledge of its own previous output state in order to determine its current state, which essentially enables the de-multiplexing of the data at the input of the receiver to allow faster data rates[11]. For example, a simple high-pass FE implemented using discrete-time techniques (with a sample-and-hold circuit at the input) could be represented by:

$$V_{Out}(Z) = \left(1 - \alpha Z^{-1}\right) V_{In}(Z) \qquad (6\text{-}6)$$

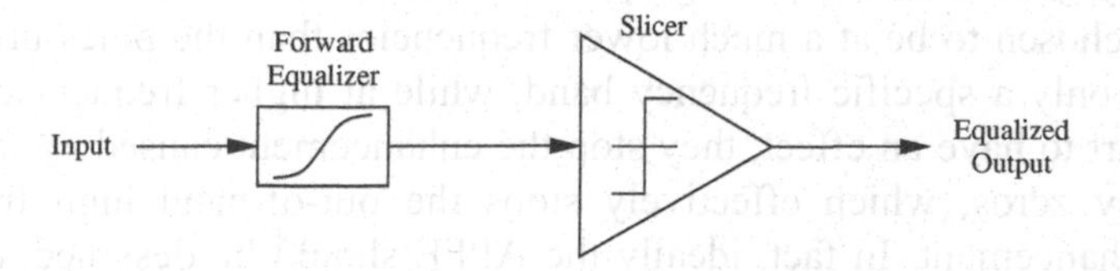

Figure 6-16. The feed-forward equalization architecture.

or in the time domain:

$$V_{Out}(t) = V_{In}(n) - \alpha V_{In}(n-1) \tag{6-7}$$

If the incoming data is sampled using n equally-spaced phases of a reference clock running at $1/n$ the data rate (Φ_1, Φ_2, ... Φ_n), then n forward equalizers (represented by Eq. 6-6) could be used in parallel, with each equalizer running at $1/n$ the data rate and using two successive phases of the reference clock to sample $V_{In}(n)$ and $V_{In}(n\text{-}1)$. For example, the first FE will use $\Phi_1$and Φ_2, while the second FE will use Φ_2, Φ_3 and so forth. Effectively, this scheme de-multiplexes the received data into n parallel lines with each line running at only $1/n$ the data rate. This essentially relaxes the design of the FE since each equalizer stage now has n bit-time to converge to its final result. In other words, much higher data rates could be equalized without requiring each equalizer to be as fast as the data rate[11].It is important to note here that other equalization architectures that rely on feedback, i.e. knowledge of previous output states is required in order to determine the current state, do not accommodate the de-multiplexing feature. In the next section, a more detailed discussion of lack of de-multiplexing capability in feed-back architectures will be presented.

The major disadvantage on the other hand of the feed-forward architecture is its relatively poor noise performance[10]. This stems from the fact that not only does the high-pass response of the forward equalizer restore the high frequency components of the received signal, it also enhances high frequency noise as well. This enhanced high frequency noise can significantly degrade the bit error rate and potentially defeat the whole purpose of removing ISI. Therefore, the feed-forward architecture is not usually an adequate choice in highly noisy environments. Even though this is the case, there exist some techniques in the literature that tries to solve the high frequency noise boosting in feed-forward architectures. One technique implements the forward equalizer as an all-pass filter instead of simply a high-pass filter. In that case, the forward equalizer is usually called an all-pass forward equalizer or APFE for short[10,12-15]. In an all-pass filter, the DC gain and the high frequency gain are the same, as opposed to a higher high-frequency gain in the simple high-pass filter case. In APFEs, the location of zeros is chosen to be at a much lower frequencies than the poles in order to enhance only a specific frequency band, while at higher frequencies, when poles start to have an effect, they stop the enhancement caused by the lower frequency zeros, which effectively stops the out-of-band high frequency noise enhancement. In fact, ideally the APFE should be designed such that when the input is white noise, the output is also white. Figure 6-17 shows the frequency response of a low-pass transmission channel, an all-pass forward

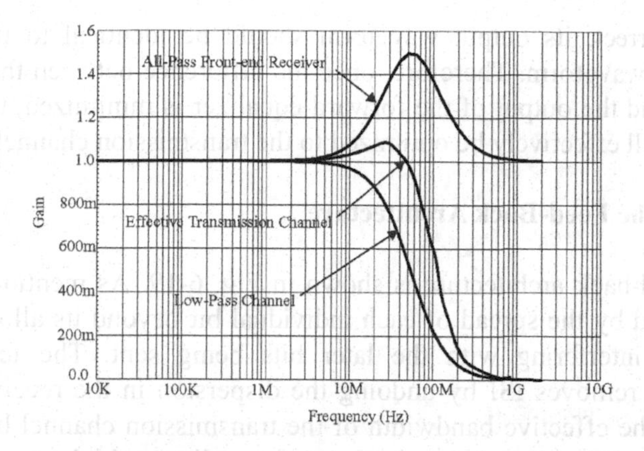

Figure 6-17. The frequency response of an all-pass forward equalizer.

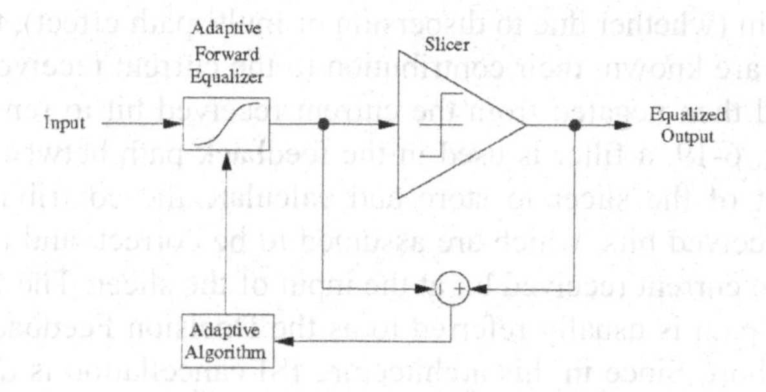

Figure 6-18. The adaptive feed-forward equalization architecture.

equalizer, and the combined response of the channel and the equalizer. While APFEs have the advantage of suppressing the high frequency noise enhancement, they add more ISI to the received signal since they simply introduce more poles to the whole transfer function of the transmission channel. So essentially, they improve the total SNR on the expense of adding more ISI.

In the feed-forward architecture shown in Fig. 6-16, the frequency response of the forward equalizer is optimized for a specific transmission channel. Therefore, it is referred to as a fixed equalizer. In order to equalize for different channels, the adaptive feed-forward architecture shown in Fig. 6-18 should be employed. In this architecture, the output of the forward equalizer is subtracted from the output of the slicer, which is assumed to have made the correct decision, to produce an error signal. This error signal is then used to change the frequency response of the forward equalizer. The theory behind this adaptive architecture is that by assuming that the output of slicer is correct, its output waveform should be identical to the original transmitted waveform. Therefore, once the difference between the output of the slicer and the output of the forward equalizer is minimized, the forward equalizer will effectively be optimized to the transmission channel.

6.3.2 The Feed-Back Architecture

The feed-back architecture is shown in Fig. 6-19. As mentioned before, ISI is caused by the spread of each individual bit beyond its allocated time, and hence interfering with the later bits being sent. The feed-forward architecture removes ISI by undoing the dispersion in the received bits by improving the effective bandwidth of the transmission channel between the transmitter and the slicer using the forward equalizer (a high-pass filter). The feed-back architecture uses a different approach. Essentially, since the current received bit has been contaminated by portions of the previous bits in the data stream (whether due to dispersion or multi-path effect), then if those previous bits are known, their contribution to the current received bit can be estimated and then negated from the current received bit to remove ISI. As shown in Fig. 6-19, a filter is used in the feedback path between the output and the input of the slicer to store and calculate the contributions of the previously received bits, which are assumed to be correct, and then subtract them from the current received bit at the input of the slicer. The filter used in the feedback path is usually referred to as the Decision Feedback Equalizer or DFE for short. Since in this architecture ISI cancellation is done through the knowledge of previously received bits, the DFE is usually called a post-cursor ISI remover[10]. The type of the DFE filter is usually dictated by the transmission channel since it has to have the same response as the channel in order to produce the same ISI effect, which in turn is being negated from the received waveform. As in the feed-forward architecture, the feed-back architecture shown in Fig. 6-19 is considered a fixed equalization architecture since the DFE is optimized for a specific channel. Figure 6-20 shows the adaptive version of the feed-back architecture. The adaptive algorithm compares the input and output waveforms of the slicer and generates an error signal. By minimizing this error signal, the DFE frequency response is adapted to the transmission channel.

The major advantage of the feed-back architecture is that it does not boost the high frequency noise as opposed to the feed-forward architecture[16]. Therefore, it generally has a better performance than the feed-forward architecture in noisy environments. The feed-back architecture is also more effective in equalizing channels with spectral nulls (such as some bad radio channels) and in combating ISI in time variant multi-path channels[3]. The disadvantage of the feed-back architecture on the other hand is that although

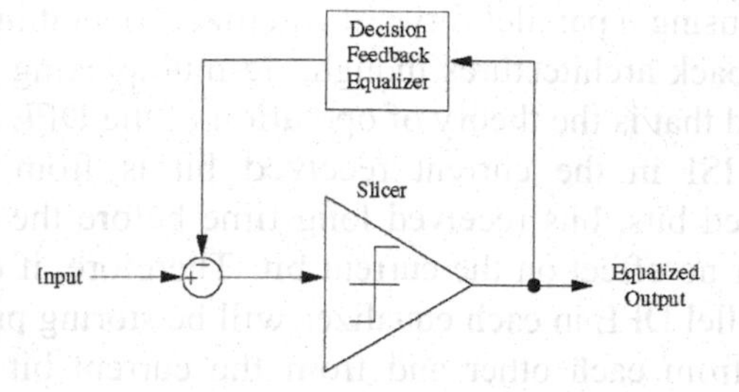

Figure 6-19. The feed-back equalization architecture.

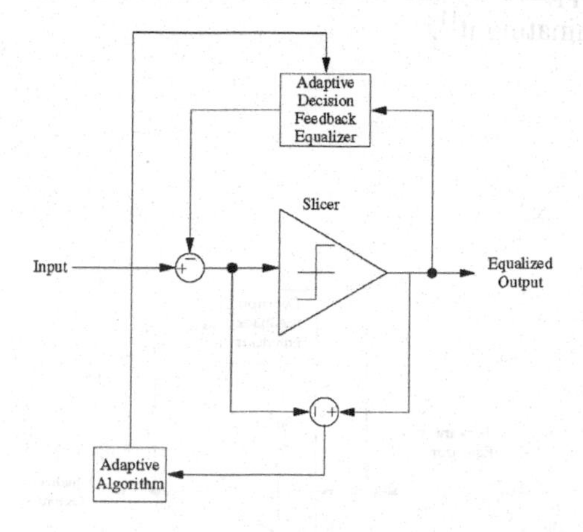

Figure 6-20. The adaptive feed-back equalization architecture.

post-cursor ISI could be removed, the waveform at the input of the slicer was not restored since the high frequency components were not boosted due to the lack of a FE (as opposed to the feed-forward architecture). This makes the design of the slicer much more difficult. Another disadvantage of the feed-forward architecture is that in order for the whole system to achieve the best performance, the total response of the channel has to be raised cosine response, which will require the use of a matched filter at the input[17]. Another disadvantage of the feed-back architecture as opposed to the feed-forward architecture is data rate limitation. The DFE function is based on calculating the ISI contributed by previously received bits in order to remove that contribution from the current bit. This means that the DFE has to finish this calculation and subtraction in a one-bit interval, which means that it has to operate as fast as the data rate. This limits the data rates the feed-back architecture can handle. As mentioned in the feed-forward architecture, this limitation could be overcome by de-multiplexing the received data to *n* parallel lines and using n parallel forward equalizers operating at $1/n$ the data rate. In the feed-back architectures though, de-multiplexing is not possible. The reason behind that is the theory of operation of the DFE itself. Since the main source of ISI in the current received bit is from the immediate previously received bits, bits received long time before the current bit will have very little or no effect on the current bit. Therefore, if de-multiplexing is used, each parallel DFE in each equalizer will be storing previous bits that are n bits apart from each other and from the current bit due to the de-multiplexing process. This means that those stored bits have very little or no contribution to the ISI in the current bit the equalizer is handling. Hence, the negated values supplied by the DFE will be incorrect and will cause more ISI instead of eliminating it[11].

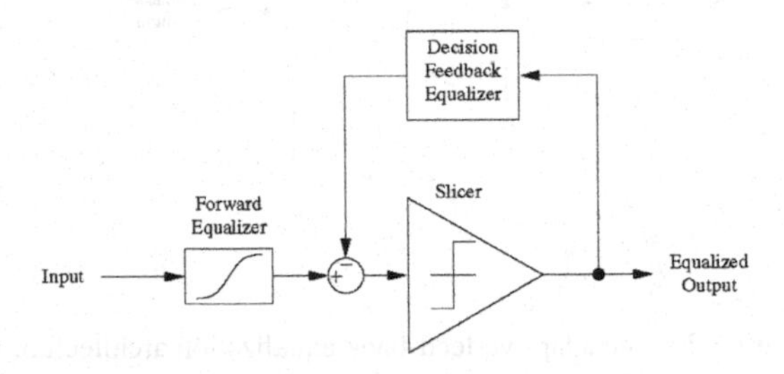

Figure 6-21. The mixed feed-forward feed-back equalization architecture.

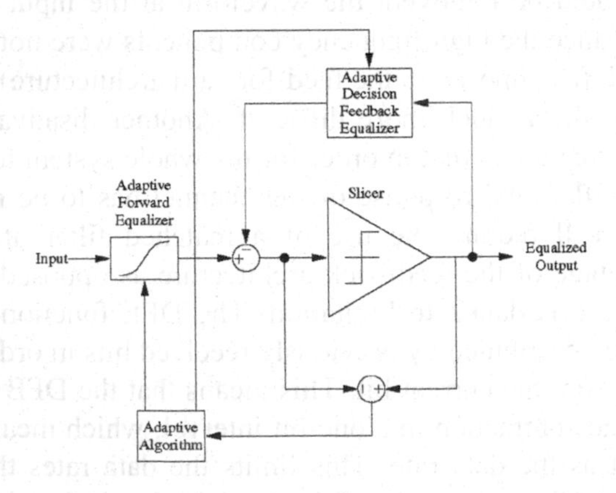

Figure 6-22. The adaptive mixed feed-forward feed-back equalization architecture.

6.3.3 The Mixed Feed-Forward Feed-back Architecture

This architecture is shown in Fig. 6-21. As shown in figure, the architecture uses both the feed-forward technique (FE) and the feed-back technique (DFE) simultaneously. Generally, this architecture is used for the severely distorted channels to achieve better equalization. The FE removes the pre-cursor ISI while the DFE removes the post-cursor ISI. It is worth mentioning that the mixed feed-forward feed-back architecture suffers from the disadvantages of both feed-forward and feed-back architectures. The adaptive version of this architecture is shown in Fig. 6-22 and it works the same way as in the two previous architectures.

6.4 Implementation Techniques

In the previous section, the different architectures used for equalization was presented. There are multiple circuit and system level implementation methodologies that could be used for realizing a given architecture. Equalization architectures and adaptive architectures in particular, heavily use programmable filters. Those filters could be implemented using purely analog techniques (continuous-time or discrete-time), or digital techniques.

Analog and digital implementations could be compared on many different levels. From a power consumption standpoint, analog implementations might potentially consume less power since there is no need for an analog to digital converter (ADC). ADCs relatively consume significant power. In addition to saving ADC power, analog implementations do not require a clock, which save the power used to generate and distribute those clocks. From an area standpoint, analog implementations are likely to consume more area even though an ADC might not be needed. This is due to the fact that signal processing in the analog domain might require potentially large circuits. For example, a low-pass filter in the analog domain requires large capacitors, while in the digital domain, it is just few flip-flops. From a difficulty standpoint, analog implementations are more difficult to design since signal processing in the analog domain is rather a very complicated task. For example, to design a filter in the analog domain, process and temperature variations, accuracy of components used, speed, stability, and noise issues become much more dominant than in the digital implementations. This makes the design in the analog domain very difficult. Digital implementations on the other hand are straight forward. For example to design a filter in the digital domain, it is a matter of calculating the filter's coefficient, then automated CAD tools are used to synthesize the circuits including layout. This makes digital implementations much more predictable and easier to design especially in digital CMOS technologies.

6.5 Summary

In this chapter, an overview of the problem of ISI in wireless and wire line digital communication systems has been presented. Eye diagrams as well as cable models were discussed due to their importance in evaluating the effect of ISI on the received bit error rate. Equalization architectures used to combat ISI have also been reviewed including feed-forward, feed-back, and mixed feed-forward feed-back architectures. The advantages and disadvantages of analog and digital implementations of those architectures from area, power, and complexity perspective were also presented.

LIST OF REFERENCES

1. U. S. H. Qureshi, "Adaptive Equalization," *Proc. IEEE* 1985.
2. Simon Haykin, "Adaptive Filter Theory," Fourth Edition, Prentice-Hall, New Jersey, 2002.
3. J. G. Proakis, "Digital Communications," Fourth Edition, McGraw-Hill, New York, 2001.
4. Herbert Taub, Donald L. Schilling, "Principles Of Communication Systems," Second Edition, McGraw-Hill, New York, 1992.
5. Bang-Sup Song, David C. Soo, "NRZ Timing Recovery Technique for Band-Limited Channels," *IEEE Journal of Solid-State Circuits*, vol. 32, NO. 4, April 1997.
6. Mohammed Ismail, Terri Fiez, "Analog VLSI Signal and Information Processing," McGraw-Hill, New York, 1994.
7. Phillip E. Allen, Douglas R. Holberg, "CMOS Analog Circuit Design," Oxford University Press, New York, 1987.
8. R. Jacob, Harry W. Li, David E. Boyce, "CMOS Circuit Design, Layout, and Simulation," IEEE Press Series on Microelectronic Systems, New York, 1998.
9. David A. Johns, Ken Martin, "Analog Integrated Circuit Design," John Wiley & Sons, New York, 1997.
10. Michael Q. Le, P. J. Hurst, J. P. Keane "An Adaptive Analog Noise-Predictive Decision-Feedback Equalizer," *IEEE Journal of Solid-State Circuits*, vol. 37, NO. 2, Feb. 2002.
11. Jae-Yoon Sim, Jang-Jin Nam, Young-Soo Sohn, Hong-June Park, Chang-Hyun Kim, Soo-In Cho "A CMOS Transceiver for DRAM BUS System with a Demultiplexed Equalization Scheme," *IEEE Journal of Solid-State Circuits*, vol. 37, NO. 2, Feb. 2002.
12. D. Lin, "High bit rate digital subscriber line transmission with noise-predictive decision-feedback equalization and block coded modulation," *Proc. IEEE Int. Conf. Communications*, pp. 17.2.1-17.3.5, 1989.
13. R. Wiedmann, J. Kenney, and W. Kolodziej "Adaptation of an all-pass equalizer for DFE," *IEEE Trans. Magn.*, vol. 35, pp. 1083-1090, Mar. 1999.
14. N. Garrido, J. Franca, and J. Kenney "A Comparative study of two adaptive continuous-time filters for decision feedback equalization read channels," *Proc. IEEE Int.Symp. Circuits and Systems*, pp. 89-92, 1997.
15. P. McEwen, and J. Kenney "All-pass forward equalizer for decision feed-back equalization," *IEEE Trans. Magn.*, vol. 31, pp. 3045-3047, Nov. 1995.
16. E. Lee, D. Messerschmitt, "Digital Communication," Kluwer, pp. 159-167, Norwell, MA 1988.
17. Simon Haykin, "Digital Communication," John Wiley & Sons, New York, 1988.

Chapter 7

AN ANALOG ADAPTIVE EQUALIZER FOR WIRE LINE TRANSCEIVERS

Intersymbol Interference is a major hold-back on the data rates that could be sent across a band limited transmission channel. As discussed in chapter 6, when sending high speed digital signals over a band limited transmission media, the high frequency components of the transmitted signal get attenuated more than the lower frequency components due to the low-pass nature of the transmission media. This unequal attenuation of the high frequency components distorts the shape of the transmitted pulse and spreads it in time beyond its allocated time slot. It also slows down the rising and falling edges of the received pulse. The time-spread of the transmitted pulse causes interference between the current received pulse and previously received ones leading the front-end receiver to make the wrong decision about the received pulse. Equalization is a technique used to solve the Intersymbol Interference problem. An Equalizer is essentially a system that compensates for the unequal attenuation caused by the transmission media, i.e. boosts the high frequency components of the received signal. An adaptive equalizer is an equalizer that can adapt its frequency response (the high frequency boosting) to transmission channels that have unpredicted frequency response. In chapter 6, different equalization architectures had been reviewed including fixed and adaptive architectures along with the pros and cons of each of them. Advantages and disadvantages of analog versus digital implementations were also discussed. In this chapter, a purely analog adaptive equalizer based on the feed-forward architecture will be presented[1]. The architecture uses a two stage tunable high-pass analog filter as its forward equalizer (FE). The FE poles locations are programmable and could be changed using a control signal. By adjusting this control signal, the frequency response of the FE could be optimized to the frequency response of the transmission channel. The adaptation to the transmission channel is achieved through comparing the edge rate of the output of the FE to a reference edge rate to generate an error signal. This error signal indicates the amount of high frequency boosting needed, and is used to generate a control signal that adjusts the frequency response of FE. The architecture does not require a training signal since it adapts itself to the channel through the edge rate of the received data rather than a specific bit pattern. The architecture has been implemented on a standard 180nm digital CMOS process with a supply range from 1.6V to 2.0V. It has been used for a wire line transceiver that operates at 125 Mbps over a variable length of up to 100 meters of an Unshielded-Twisted-Pair (UTP) Category-5 (CAT-5) Ethernet cable.

7.1 Motivation

The motivation behind this work emerges from the need of a solid, yet cheap solution for home networking. Even though wire line and wireless Ethernet networks provided a powerful solution for data networking through homes, there has been a recent increase in the demand of a home network that can handle real-time video and audio applications in addition to data. The ultimate goal of a home network is be able to have access/control over all audio, video and data equipments in the whole house from any location and to any location in the house through a single cable/connector that can handle all different types of data. The great advantage of Ethernet network is its ability to receive and transmit data at up to 100 Mbps across the super-cheap UTP cable for lengths up to 100m. This makes the Ethernet solution very appealing to home owners due to the cheap cost of wiring the whole house with UTP cables. However, Ethernet protocols were not designed to effectively handle audio/video data streams. The reason behind that is that on the hardware level, Ethernet protocols do not handle isochronous data streams. Therefore, it does not guarantee a certain bandwidth to a specific application. That could be acceptable for data transmission, but for real-time audio/video streams, this might not be practical since they require a guaranteed bandwidth. There are different protocols that have been developed to handle isochronous data streams, like IEEE 1394 and USB standards[2-4]. While those standards can handle real-time audio/video traffic, they suffer from two major disadvantages. First, they require relatively expensive shielded cables for their data transmission. Therefore, wiring the whole house could be relatively expensive. Second, the limited cable length they can handle. For example, IEEE 1394 and USB transceivers can transmit and receive over only 5 meters of cable. Since wiring a house will definitely require longer cables, using IEEE 1394 or USB transceivers is not possible for home networking. The work presented in this chapter was developed to enable high-speed transceivers in general to transmit and receive over UTP

CAT-5 cables (which is used for Ethernet networks). The presented work enables using wire line communication standards that are more adequate for real-time audio/video applications over the cheap UTP cable that is used for home networks. In order to achieve that, adaptive equalization was essential due to variability in cable length, which could be as long as 100 meters.

7.2 Transmission Channel Modeling

The first step in any equalization process is the estimation of the transmission channel behavior. This step is very important in determining the amount of equalization needed, i.e. the amount and frequency range of boosting the FE need to provide. While adaptive equalization architectures should automatically adapt to the transmission channel, it is essential to have some sort of worst-case and best-case channel model in order to define the required range of equalization. In wireless channels, this could be a very difficult task since wireless channels are subject to many different factors as mentioned before. In wire line channels on the other hand, the modeling process is relatively much easier. This is due to the fact that most wire line channels (cables) are regulated by specific standards that manufactures have to abide to. This makes the prediction of the channel's behavior much simpler. The most commonly used model for wire line channels is the lumped RLC transmission line model[5,6]. Even though cable models based on S-parameter measurements result in a much more accurate modeling of the channel, still, RLC models are much easier to use and simulate with acceptable levels of accuracy. For the equalizer implementation, the three section RLC transmission line model shown in Fig. 7-1 was used to model a 100 meter UTP CAT-5 cable. This model introduces worst-case ISI the equalizer has to deal with. In general though, it is well known that when transmitting data across long cables, the DC component of the transmitted signal has to be blocked to avoid any significant ground shifting from causing any safety hazards. Additionally, a termination network that matches the characteristic impedance of the cable (100Ω differential) has to be used to avoid any reflections on the line. For that purpose, the whole transmission channel including the cable model shown in Fig. 7-1 is shown in Fig. 7-2. It's also worth mentioning that for very short cables, the model shown in Fig. 7-2 can be used, with an acceptable accuracy, by removing the cable model and just leaving the termination networks. From now on, this will be referred to as the short-cable model, or the best-case transmission channel.

In order to assess the behavior of the transmission channel, Fig. 7-3 shows the frequency response of the transmission channel shown in Fig. 7-2 for both the 100m and short-cable cases. The bandwidth of the 100m channel is 1.75 MHz, which is almost 100 times less than the targeted data rate of 125 Mbps, while the short-cable channel bandwidth is 280 MHz. In the time domain, eye diagrams are usually the best tool to asses the effect of the limited bandwidth of the transmission channel on the transmitted data. Figures 7-4 and 7-5 show the eye diagrams of a random data stream at the input and the output of a 100m transmission channel. As shown in those figures, the input eye is ideal, while the output eye is completely closed. Hence, equalization is necessary to recover the transmitted data correctly.

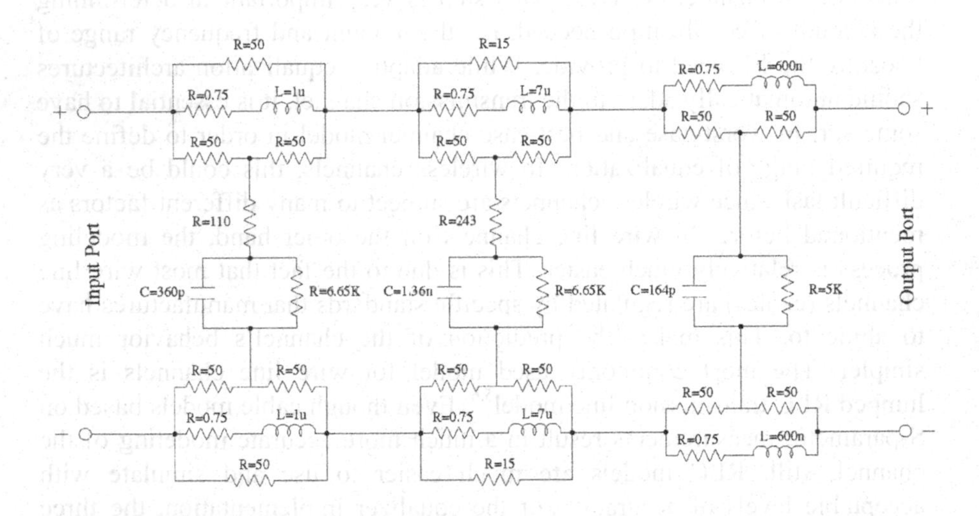

Figure 7-1. An RLC transmission line model for 100m UTP CAT-5 cable.

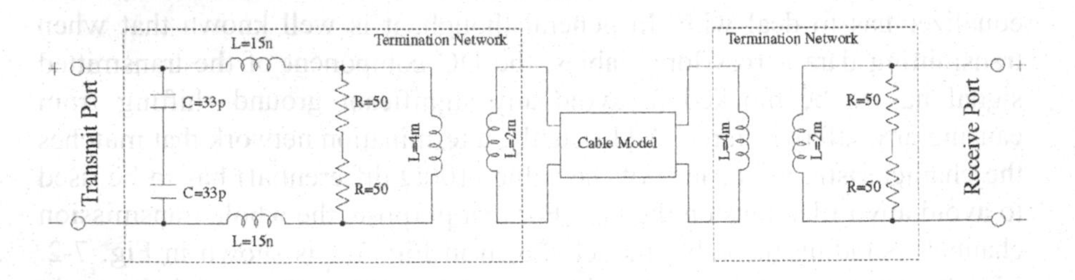

Figure 7-2. The transmission channel including termination networks.

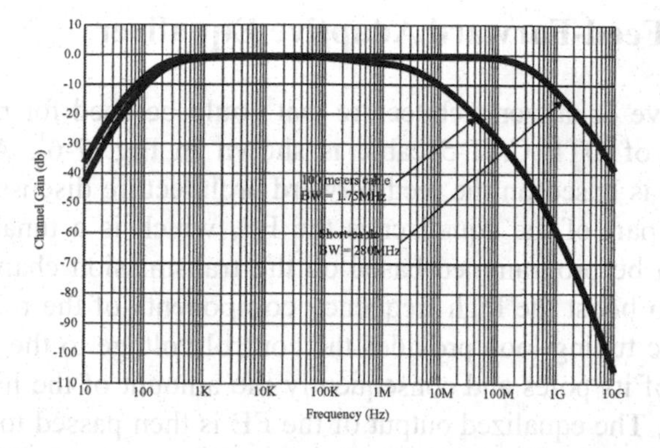

Figure 7-3. **The frequency response of 100m, and short-cable transmission channels.**

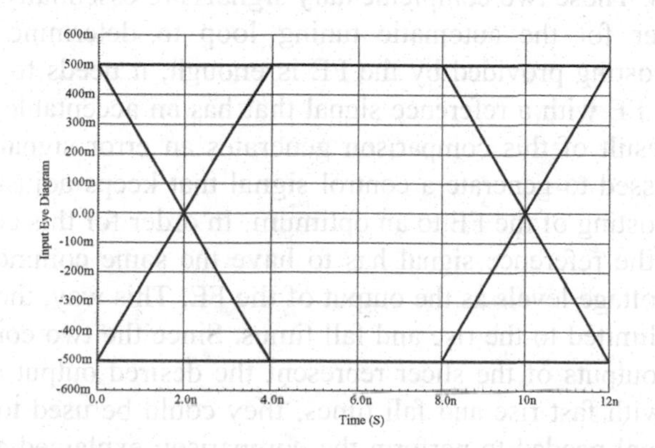

Figure 7-4. **Eye diagram of the 100m transmission channel input**

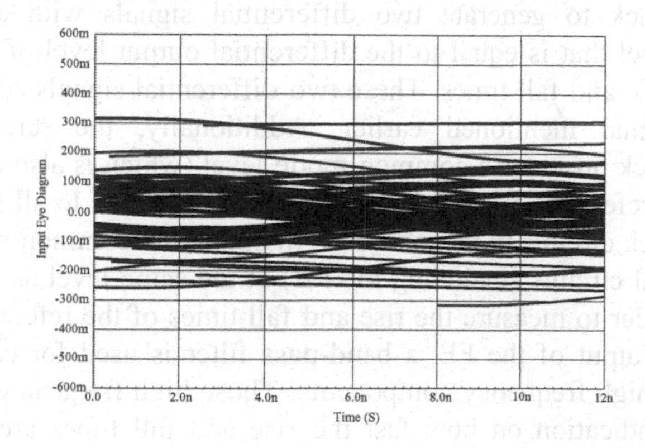

Figure 7-5. **Eye diagram of the 100m transmission channel output.**

7.3 A Feed-Forward Adaptive Equalizer

An adaptive equalizer architecture that could be used for receiving data across 100m of UTP CAT-5 cable is shown in Fig. 7-6[1]. As mentioned previously, it is based on the feed-forward architecture discussed in chapter 6. The main part of the equalizer is the FE, which is a tunable high-pass filter that can be programmed based on the transmission channel behavior, and is used to boost the high frequency components of the received signal. The automatic tuning loop provides the control voltage to the FE to control the location of its poles and consequently the amount of the high frequency enhancement. The equalized output of the FE is then passed to a high-speed comparator (the slicer), which resolves the differential signal output of the FE and converts it to two fully digital complementary signals with fast rise and fall times. Those two complementary signals are essentially the received data. In order for the automatic tuning loop to determine if the high frequency boosting provided by the FE is enough, it needs to compare the output of the FE with a reference signal that has an acceptable rise and fall times. The result of this comparison generates an error signal that can be further processed to generate a control signal that keeps adjusting the high frequency boosting of the FE to an optimum. In order for this comparison to be accurate, the reference signal has to have the same common-mode and differential voltage levels as the output of the FE. This way, the comparison will be only limited to the rise and fall times. Since the two complementary fully digital outputs of the slicer represent the desired output of the whole system, i.e. with fast rise and fall times, they could be used to provide the reference signal needed to perform the comparison explained above. These two signals though are digital rail to rail signals, and therefore they can't be used directly. Instead, they are used in conjunction with a reference signal generator block to generate two differential signals with a differential amplitude level that is equal to the differential output level of the FE, only with faster rise and fall times. These two differential signals are used as the reference signal mentioned earlier. Additionally, the reference signal generator block provides a common-mode level (which is also the common-mode of the reference signal) that is used as a reference to all the common-mode feedback circuits in the system to maintain the common-mode level of all differential circuits (including the FE) at the same level as the reference signal. In order to measure the rise and fall times of the reference signal as well as the output of the FE, a band-pass filter is used for each signal to measure the high frequency components. Those high frequency components provide an indication on how fast the rise and fall times are, or in other words, it is like measuring the slopes of the signals. Since the parameter of interest here is the edge rate rather than its direction (rising or falling edge),

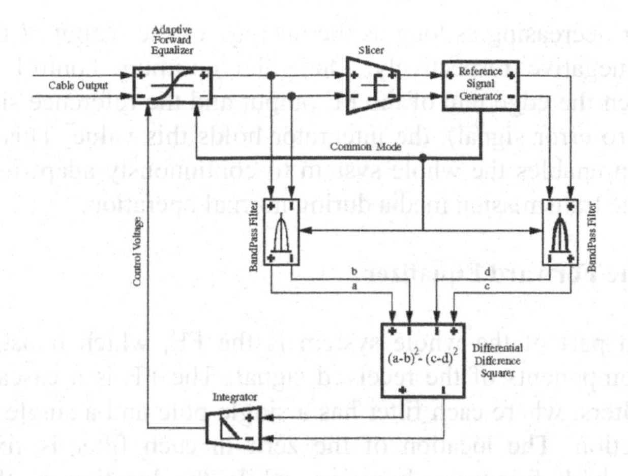

Figure 7-6. Adaptive equalizer block diagram.

the output of each band-pass filter is squared. The reason behind that is that the band-pass filter yields a positive output for a rising edge and a negative output for a falling edge. By squaring the band-pass filter's output, a correct indication of the edge rate could be estimated regardless of the direction of the edge. The squared outputs of the two band-pass filters are then subtracted to generate an error signal. The sign of this error signal indicates whether the FE is over-boosting or under-boosting the high frequency components of the received signal. Therefore, if the rising and falling edges of the FE output are slower than the reference signal, the error signal will have a positive sign, which indicates under-boosting. On the other hand, if the rising and falling edges of the forward equalizer output are faster than the reference signal, the error signal will have a negative sign, which indicates over-boosting. Optimum boosting will be reached once the error signal converges to zero. The Differential Difference Squarer (DDS) block shown in Fig. 7-6 performs the squaring and the subtracting of the outputs of the two band-pass filters to generate the error signal. Since the output signals of the band-pass filters will only have energy during transitions in the input signals and zero otherwise, the DDS output will also have energy during transitions only. This means that the error signal will have very high frequency components that could potentially cause instability of the whole adaptive loop if it was used directly to control the FE. In order to remove those high frequency components from the error signal, an integrator is used as a low-pass filter to remove the high frequency components, and to also average the error signal over time. The integration process results in a smoothly changing signal that can be used to control the frequency response of the FE. The output of the integrator (the control signal) will keep increasing or decreasing as long as the average of the output of the DDS is positive or negative respectively. Once the optimum control voltage is reached (when the edge rate of the FE output and the reference signal is the same, i.e. zero error signal), the integrator holds this value. This automatic adaptive loop enables the whole system to continuously adapt itself to any changes in the transmission media during normal operation.

7.3.1 The Forward Equalizer

The main part of the whole system is the FE, which boosts the high frequency components of the received signal. The FE is a cascade of two first-order filters, where each filter has a single pole and a single zero in its transfer function. The location of the zero in each filter is fixed and it provides the high frequency boosting, while the location of the pole is adjustable using a control voltage. By increasing the control voltage, the position of the pole in each filter shifts to higher frequencies allowing more boosting (introduced by the zero in the transfer function), while by decreasing the control voltage, the location of the pole in each filter shifts to lower frequencies to reduce the boosting. In all cases, the location of the pole in each filter is always kept at a higher frequency than the zero in order to allow boosting. The FE design uses two cascaded fully differential first-order G_M-C filters. In each filter stage, two identical voltage-controlled transconductors, two matched resistor, and two matched capacitors are used to implement the filter. Since the design is fully differential, a common-mode feedback circuit that keeps the common-mode level at a constant value is used. Figure 7-7 shows the circuit diagram of the first-order G_M-C filter used. The FE uses two cascaded stages of this G_M-C stage. The transfer function of each filtering stage could be written as:

$$\frac{V_O}{V_i} = \frac{G_m}{g} \frac{(s \times 2C + g)}{s \times C + G_m} \tag{7-1}$$

where G_m is the transconductance of the transconductor stage, while $g = 1/R$. As Eq. 7-1 shows, the pole location is controlled by the transconductance value G_m, which is in turn controlled by the control voltage. The zero location in the first stage of the FE is designed to be at 3.32 MHz nominally, with $R = 12\ \text{K}\Omega$ and $C = 4$ pF, while in the second stage it is designed to be at 18.95 MHz with $R = 21\ \text{K}\Omega$ and $C = 0.4$ pF. The common-mode feedback circuit tracks the common-mode voltage of the output of the fully differential G_M-C stage and compares it to a set common-mode level provided by the reference signal generator. If the output common-mode is

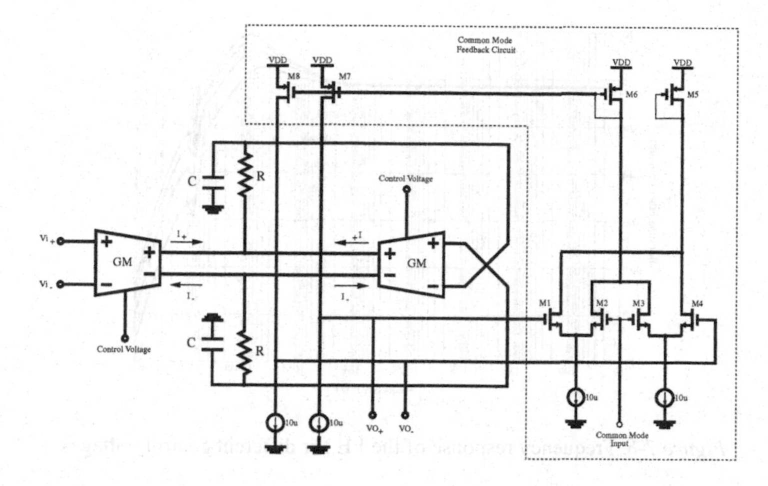

Figure 7-7. First-order G_M-C high-pass filter with common-mode feedback.

equal to the required common-mode, the drain current of M_6 will be equal to the tail current of the differential pairs. Because M_7 and M_8 are identical to M_6, both of them will source that same current into the two single-ended outputs of the filter, but since there are two current sinks with the same value connected to each single-ended output, the common-mode will stay the same. On the other hand, if the common-mode voltage of the output was higher than the required common-mode, the drain current of M_6 will decrease, leading to the reduction of the currents sourced by M_7 and M_8, which will cause the two current sinks connected to the outputs to pull the common-mode voltage down until it gets to the desired value. The same logic applies if the common-mode of the output of the filter was lower than the desired common-mode.

The voltage-controlled transconductor is an essential element to the implementation of the FE. As discussed in chapter 2, there are many techniques reported in the literature for implementing voltage controlled transconductors[7,8]. In chapter 3, a voltage-controlled transconductor was presented in detail and is used here to implement the G_M-C stages of the FE. Figure 7-8 shows the frequency response of the FE for different control voltage. Note that the supply voltage is 1.8V. As shown in figure, by increasing the control voltage, the high frequency boosting will also increase. It is worth mentioning that the resistors used in the FE are n-well resistors. The reason of choosing n-well resistors over silicide-block resistors is the positive temperature coefficient of the well resistors. As shown in Eq. 7-1, the high frequency gain of the FE is equal to $G_m \times R$, and since G_m has a

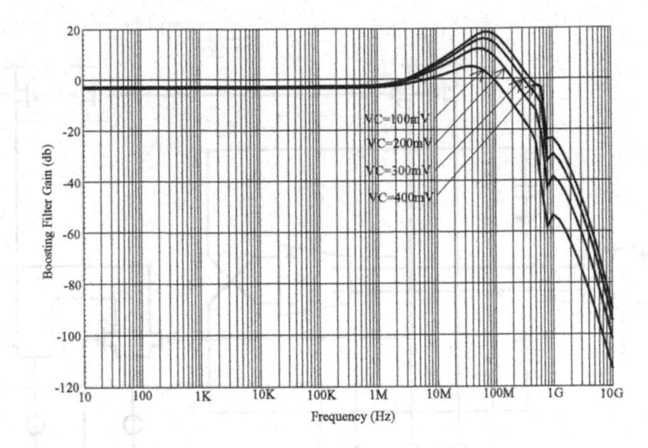

Figure 7-8. Frequency response of the FE for different control voltages.

negative temperature coefficient (due to the degradation of the input transistors transconductance with temperature), having positive temperature coefficient resistors compensates for the negative temperature coefficient of G_m, and hence achieves a stable high frequency gain with temperature.

7.3.2 The Slicer

The slicer, which is simply a comparator, is the circuit element that resolves the differential output of the FE. Its bandwidth has to be enough to handle a data rate of 125 Mbps, and it serves two purposes in the system. First, it resolves the output of the FE and generates two rail-to-rail complementary digital signals that serve as the received data. Second, its output is used by the reference signal generator to generate the differential reference signal needed by the system to adapt the FE, as well as a common-mode reference used by all common-mode feedback circuits in the system. Figure 7-9 shows the schematic diagram of the slicer, which has three amplification stages. The first stage is a current mirror operational transconductance amplifier (OTA) that gives a moderate amplification for the input signal and provides a buffering between the input and the output to prevent kick-back[7]. Transistors M_{13}, and M_{14} have a very important role in the circuit, they prevent M_3 and M_4 from completely turning off during slew rate limited operation. That way, the circuit will not need to recharge the gates of M_3 and M_4, which significantly increases the speed of the slicer. The second stage of the slicer is a self-biased differential amplifier for each complementary output. This second stage is used to provide more amplification to the signal and to perform the differential to single-ended conversion of the received signal.

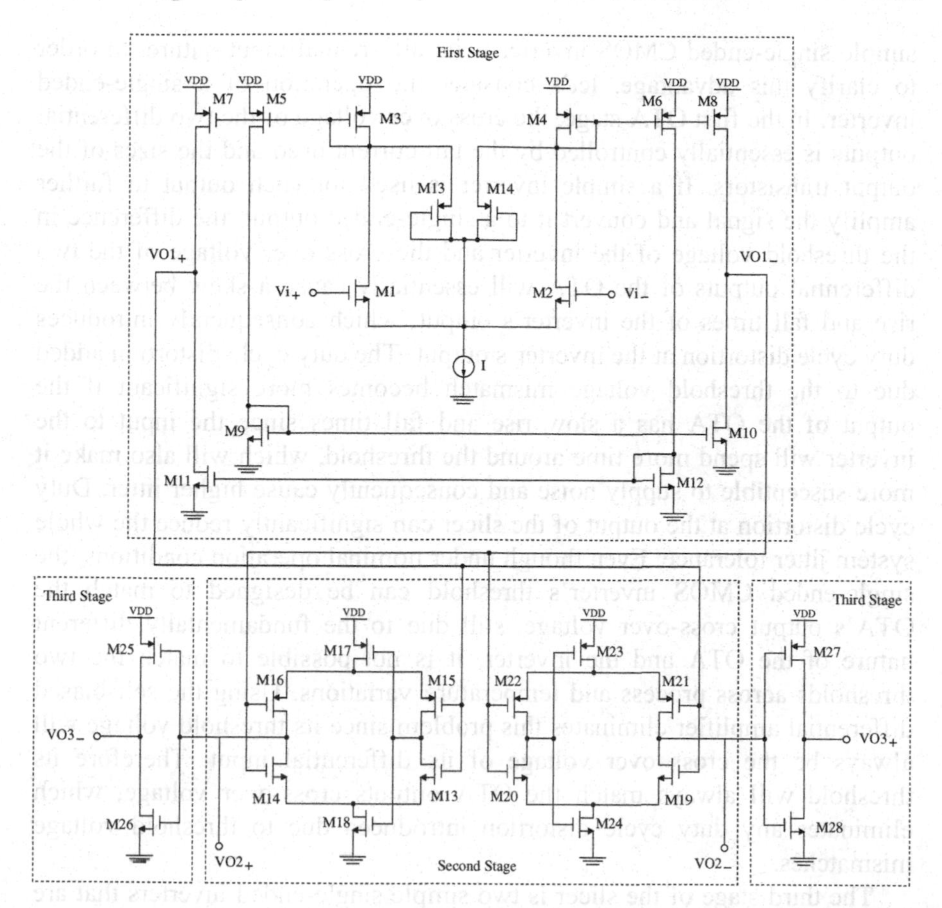

Figure 7-9. Schematic diagram of the three-stage slicer.

Using the self-biased differential amplifier for differential to single-ended conversion has many advantages over using just a simple single-ended inverter, or the classic constant-tail-current differential amplifier[9]. First, as opposed to the classic differential amplifier, the self-biased differential amplifier is significantly faster during switching since, due to local feedback, it can supply a relatively high current to speed up the rise and fall times of the transition. In a classic differential amplifier on the other hand, the speed is always limited to the value of the tail current used. Another advantage of the self-biased differential amplifier over the classic differential amplifier is that for digital signals, the self biased differential amplifier consumes current only during transitions, while in a classic differential amplifier, the tail current will be always consumed even if there were no transitions. The second advantage of the self-biased differential amplifier as opposed to a simple single-ended CMOS inverter is its differential input nature. In order to clarify this advantage, let's consider the operation of a single-ended inverter. In the first OTA stage, the cross over voltage of the two differential outputs is essentially controlled by the tail current used and the sizes of the output transistors. If a simple inverter is used for each output to further amplify the signal and convert it to a single-ended output, the difference in the threshold voltage of the inverter and the cross over voltage of the two differential outputs of the OTA will essentially cause a skew between the rise and fall times of the inverter's output, which consequently introduces duty cycle distortion at the inverter's output. The duty cycle distortion added due to the threshold voltage mismatch becomes more significant if the output of the OTA has a slow rise and fall times since the input to the inverter will spend more time around the threshold, which will also make it more susceptible to supply noise and consequently cause higher jitter. Duty cycle distortion at the output of the slicer can significantly reduce the whole system jitter tolerance. Even though under nominal operation conditions, the single-ended CMOS inverter's threshold can be designed to match the OTA's output cross-over voltage, still due to the fundamentally different nature of the OTA and the inverter, it is not possible to match the two thresholds across process and temperature variations. Using the self-biased differential amplifier eliminates this problem since its threshold voltage will always be the cross over voltage of its differential input. Therefore its threshold will always match the OTA outputs cross-over voltage, which eliminates any duty cycle distortion introduced due to threshold voltage mismatches.

The third stage of the slicer is two simple single-ended inverters that are used to further increase the gain of the whole slicer. The reason using single-ended inverters at this point is acceptable, while using them right after the OTA is not, is that the outputs of the self-biased differential amplifier have significantly faster rise and fall times than the OTA outputs. This faster rise and fall times minimizes the duty cycle distortion added due to the threshold voltage mismatch between the self-biased differential amplifier and the single-ended inverter. Figure 7-10 shows the frequency response of the first stage of the slicer. Since the second and third stages are only for the differential to single-ended conversion, the speed of the whole slicer will be limited by the first stage. As shown in Fig. 7-10, the bandwidth of the first stage is 21 MHz, while the unity gain frequency is at 1.96 GHz. The DC gain of the slicer is 37 dbs.

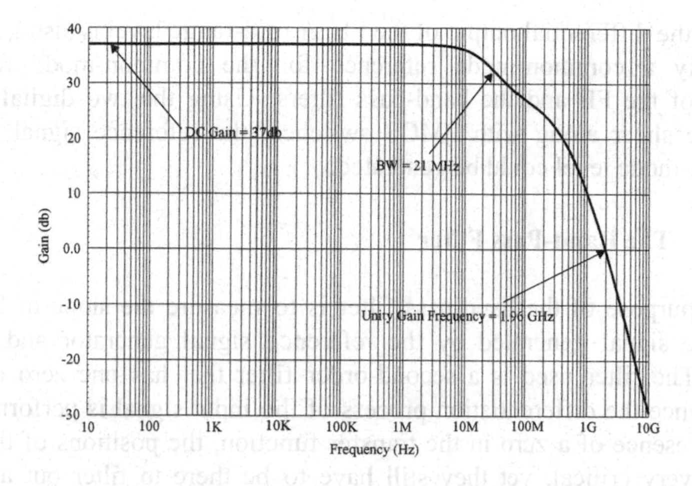

Figure 7-10. Frequency response of the slicer first stage.

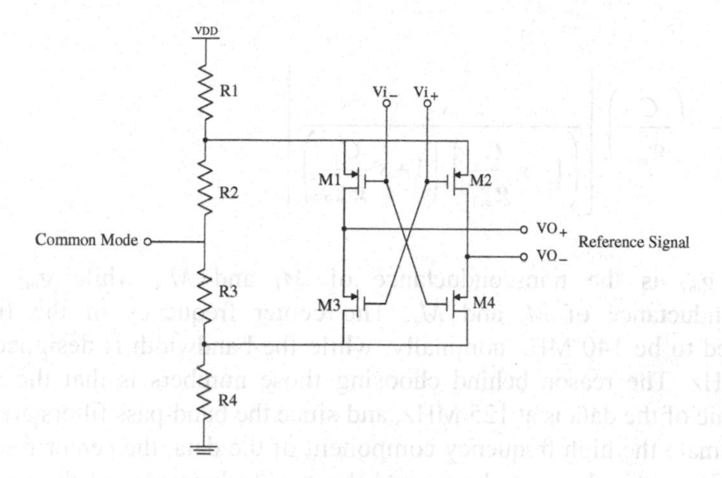

Figure 7-11. Reference signal generator schematic diagram.

7.3.3 The Reference Signal Generator

The slicer provides two complementary rail-to-rail digital signals that can be used to generate the reference signal needed for the comparison process with the FE output. As mentioned before, this reference signal has to have the same common-mode and differential amplitude as the FE output. In order to achieve that, a potential divider is used to generate two voltage levels as shown in Fig. 7-11. The difference between these two levels is equal to the differential output of the FE. A mid-range level is also generated to supply a common-mode reference for the common-mode feedback circuits of the FE and the band-pass filters. Using the two digital signals from the slicer along with PMOS switches, the reference signal and the common-mode level could be generated.

7.3.4 The Band-Pass Filter

The purpose of the band-pass filter is to measure the slope of both the reference signal generated by the reference signal generator and the FE output. The filter used is a second-order filter that has one zero and two poles. Since the differentiation process of the input signal is performed due to the presence of a zero in the transfer function, the positions of the poles are not very critical, yet they still have to be there to filter out any high frequency noise. Figure 7-12 shows the schematic diagram of the band-pass filter. The transfer function of the filter could be written as:

$$\frac{V_{od}}{V_{id}} = \left(\frac{C}{g_{m1}}\right)\left[\frac{s}{\left(1+s\frac{C}{g_{m1}}\right)\left(1+s\frac{C_1}{g_{m2}}\right)}\right] \quad (7\text{-}2)$$

where g_{m1} is the transconductance of M_1 and M_2, while g_{m2} is the transconductance of M_3 and M_4. The center frequency of the filter is designed to be 140 MHz nominally, while the bandwidth is designed to be 200 MHz. The reason behind choosing those numbers is that the second harmonic of the data is at 125 MHz, and since the band-pass filters are trying the estimate the high frequency component of the data, the center frequency of the filter was chosen to be around the second harmonic of the data. It is worth mentioning that since the design is fully differential, a common-mode feedback circuit is used to set the common-mode level of the filters outputs to the same common-mode as the reference signal and the FE output. Figure 7-13 shows the frequency response of the filter.

7.3.5 The Differential-Difference Squarer

The main function of this block is to square two differential signals and then subtract them. It is used to compare the differential outputs of the two band-pass filters discussed in the previous section to generate the error signal. The operation of the circuit is based on the large signal model of the MOS transistor provided that the common-mode voltage of the two differential signals is the same[8]. Figure 7-14 shows the schematic diagram of the DDS. The differential signals applied to the two transistor pairs M_3, M_4 and M_5, M_6 generate the following two currents:

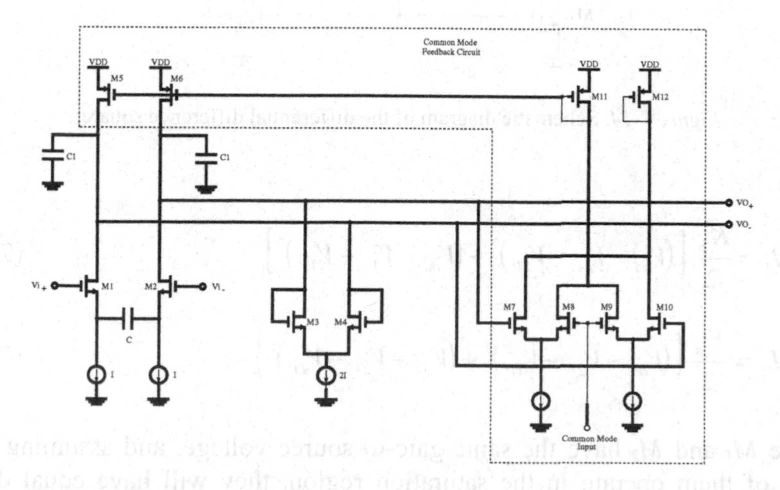

Figure 7-12. Schematic diagram of the band-pass filter.

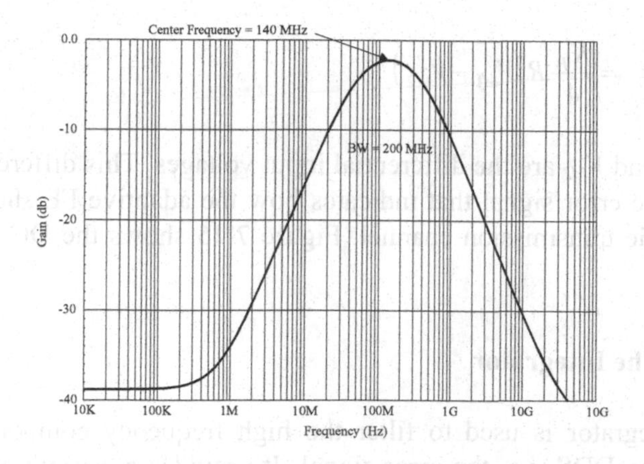

Figure 7-13. Frequency response of the band-pass filter.

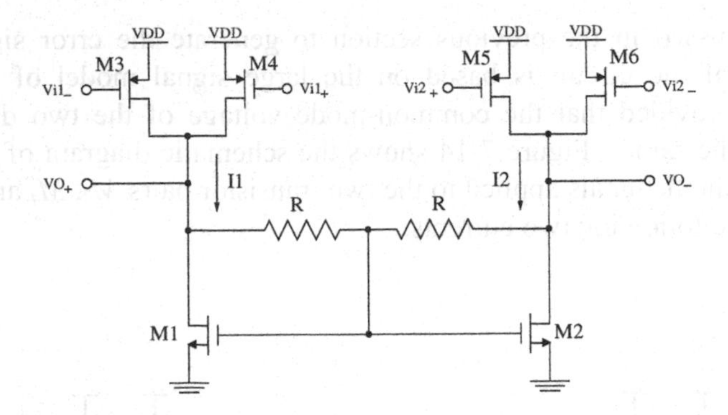

Figure 7-14. Schematic diagram of the differential difference squarer.

$$I_1 = \frac{K_p}{2}\left[\left(V_{dd} - V_{1+} - V_{TP}\right)^2 + \left(V_{dd} - V_{1-} - V_{TP}\right)^2\right] \quad (7\text{-}3)$$

$$I_2 = \frac{K_p}{2}\left[\left(V_{dd} - V_{2+} - V_{TP}\right)^2 + \left(V_{dd} - V_{2-} - V_{TP}\right)^2\right] \quad (7\text{-}4)$$

Since M_1 and M_2 have the same gate-to-source voltage, and assuming that both of them operate in the saturation region, they will have equal drain currents. Therefore, the difference between I_1 and I_2 will flow into the resistors. The differential output voltage could then be written as:

$$V_{O+} - V_{O-} = \frac{K_P}{4} R\left(V_{id1}^2 - V_{id2}^2\right) \quad (7\text{-}5)$$

where V_{id1} and V_{id2} are the differential input voltages. This differential output serves as the error signal that indicates how the adaptive FE should behave for a specific transmission channel. Figure 7-15 shows the DC response of the circuit.

7.3.6 The Integrator

The Integrator is used to filter the high frequency components in the output of the DDS, i.e. the error signal. It provides a smooth averaged-out

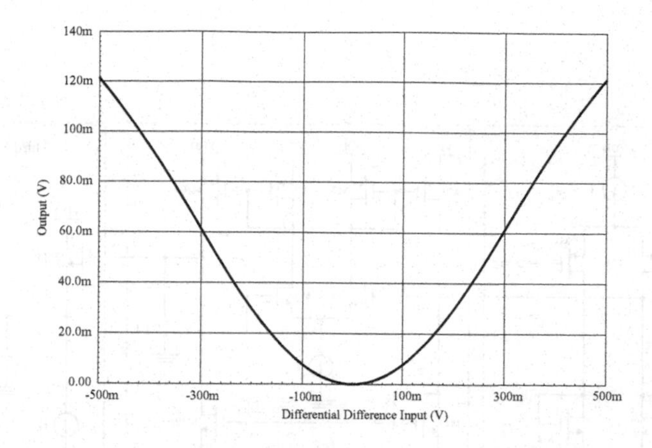

Figure 7-15. DC response of the differential difference squarer.

version of the error signal that can be used as a control signal for the FE. This filtering process is essential to avoid any instability in the whole adaptive loop, as well as to guarantee that the adaptive loop will not respond except for a real change in the transmission channel. The integrator has three stages as shown in Fig. 7-16. The first stage is a simple pre-amplifier to isolate the input and the output of the integrator and provide some amplification to the very small output of the DDS. The second stage is a single-ended OTA with a capacitive load to achieve the integration function. The output of the second stage could be written as:

$$V_O = \frac{g_m}{s\,C_1} V_{id} \tag{7-6}$$

where g_m is the transconductance of M_5 and M_6. The third stage of the integrator is a simple level shifter to shift the dc level of the integrator's output to match the control range of the FE. This shifted-down version of the integrated signal serves as the control voltage to the FE. Figure 7-17 shows the frequency response of the integrator.

In the previous few sections, the circuit design of each block in the system has been introduced. In the next section, the system level behavior will be investigated and simulation and lab measurements will be compared in order to provide more insight into the system behavior.

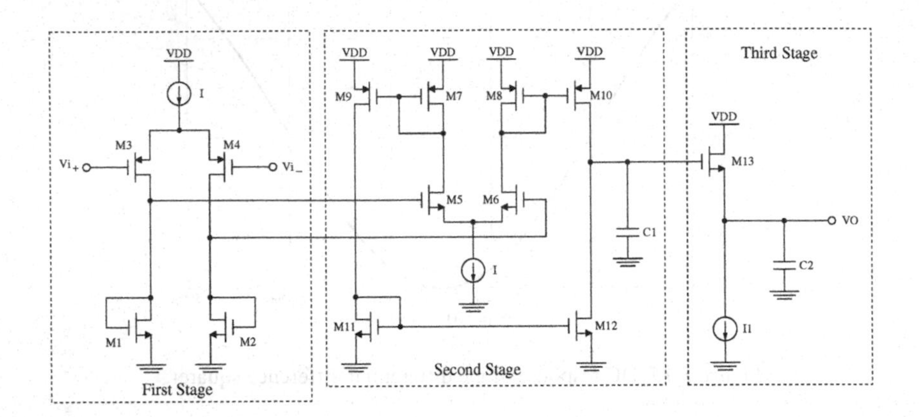

Figure 7-16. Schematic diagram of the integrator.

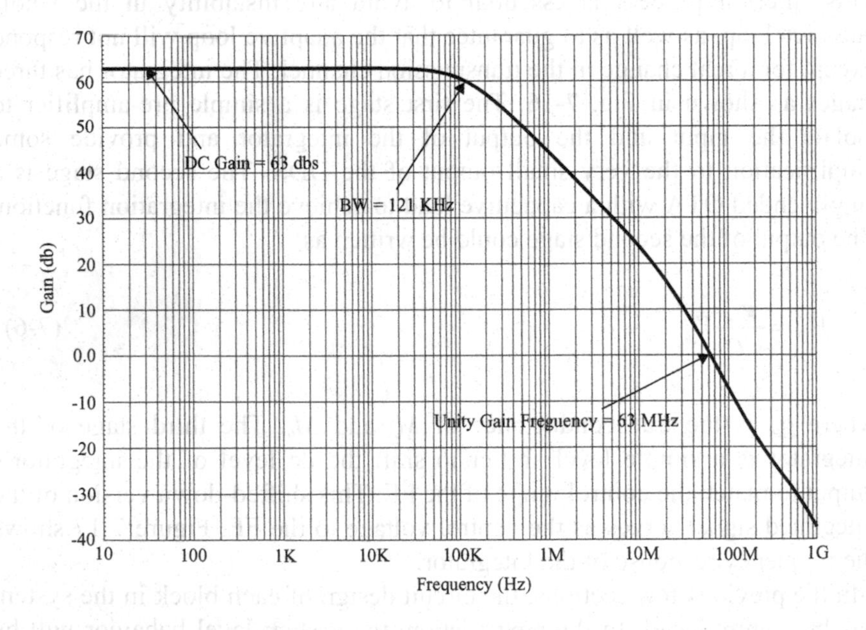

Figure 7-17. Frequency response of the integrator.

7.4 Simulation Results

The presented equalizer was implemented using a standard 180nm digital CMOS process with 1.8V supply voltage. In order to evaluate the performance of the system, two types of simulations were performed. First, an AC simulation of the FE with different transmission channels to evaluate the adaptive behavior of the equalizer. Second, a transient simulation with different transmission channels to evaluate the performance of the rest of the blocks in the system. Two extreme transmission channels were used for the purpose of simulation, the first one is the 100m cable transmission channel, and the second one is the short-cable transmission channel. Both channels were discussed in detail in section 7.2.

AC simulations were performed on the adaptive FE and the results are shown in Figs. 7-18 and 7-19. Those figures show the frequency response of the transmission channel and the FE individually, as well as their combined frequency response. As shown in Fig. 7-18, the bandwidth of the 100m channel is only 2 MHz while the data rate is 125 Mbps. Since the bandwidth of the channel is very limited the adaptive FE provides the maximum high frequency boosting as shown in figure. Therefore, the combined frequency response of the channel and the equalizer (which is the effective bandwidth) has an improved bandwidth that enables the reception of 125 Mbps data rate.

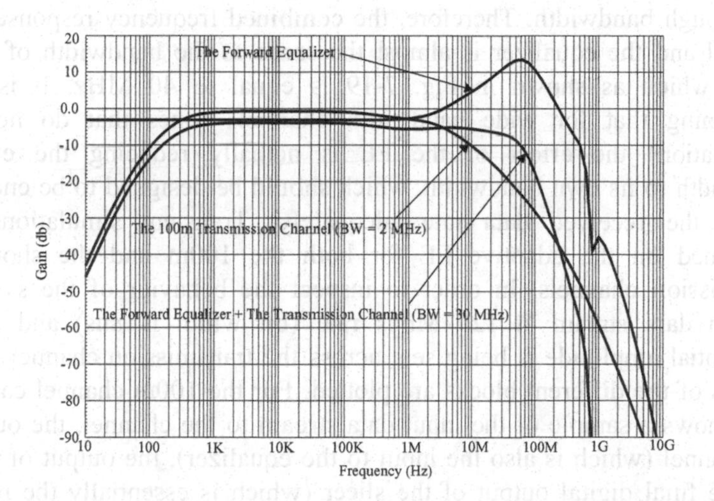

Figure 7-18. Frequency response of the FE, the 100m transmission channel, and the combined frequency response of the channel and the equalizer.

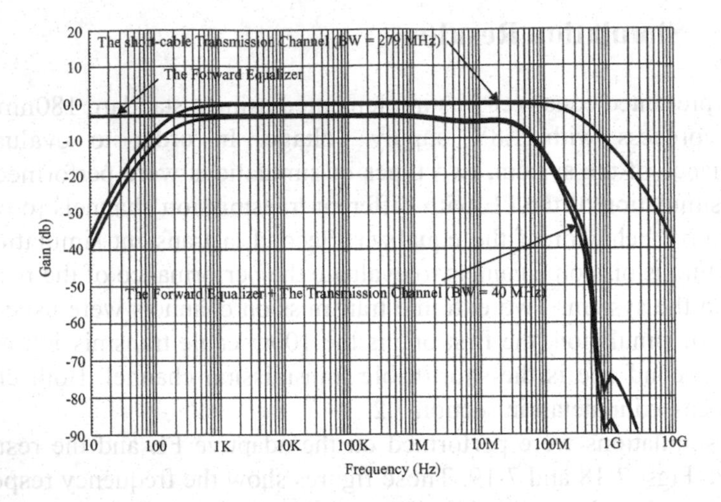

Figure 7-19. Frequency response of the FE, the short-cable transmission channel, and the combined frequency response of the channel and the equalizer.

As shown in Fig. 7-18, the effective bandwidth is 30 MHz, which is 15 times higher than the original bandwidth of the transmission channel. Figure 7-19 on other hand shows the short-cable channel case. In this case, the FE provides almost no high frequency boosting since the transmission channel has enough bandwidth. Therefore, the combined frequency response of the channel and the equalizer is almost the same as the bandwidth of the FE alone, which as shown in Fig. 7-19 is equal to 40 MHz. It is worth mentioning that for wide bandwidth channels (ones that do not need equalization) the effect of the FE is actually reducing the effective bandwidth to its own bandwidth, which should be designed to be enough to handle the received data rate adequately. Transient simulations were performed on the adaptive FE for both the 100m and the short-cable transmission channels. In order to inspect the behavior of the system, a random data stream at 125 Mbps rate (bit width is 8ns) and 500mV differential amplitude is being sent across the transmission channel and the outputs of the different blocks are plotted. For the 100m channel case, Fig. 7-20 shows a sample of the input data stream to the channel, the output of the channel (which is also the input to the equalizer), the output of the FE, and the final digital output of the slicer (which is essentially the received data). As shown in Fig. 7-20, the single-ended outputs of the transmission channel do not cross over due to the severe ISI introduced by the limited bandwidth of the channel. The single-ended outputs of the FE on the other hand cross over normally due to the high frequency boosting introduced by the FE, which effectively removed ISI caused by the channel from the received data. The output of the FE is then passed to the slicer to generate the final output. As mentioned earlier, eye diagrams are useful for evaluating equalizers performance by inspecting their vertical and horizontal openings.

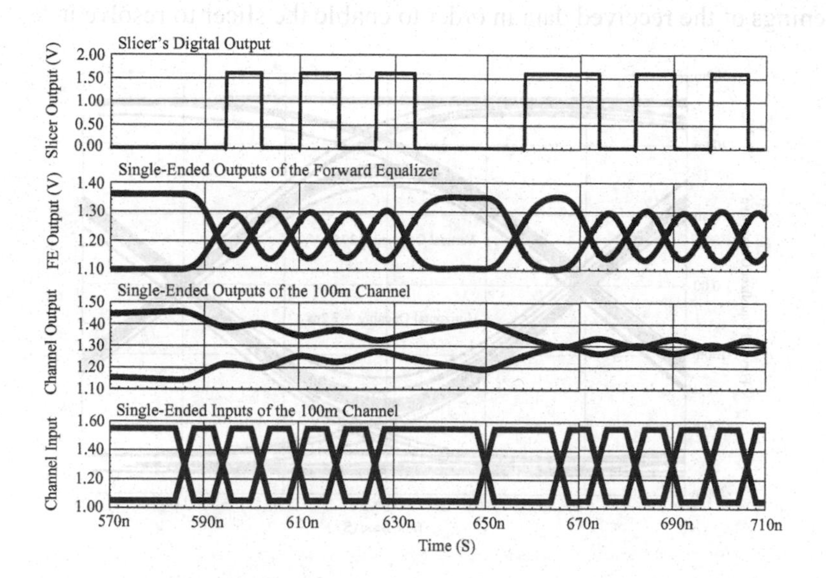

Figure 7-20. Transient response of the 100m channel, the FE, and the slicer.

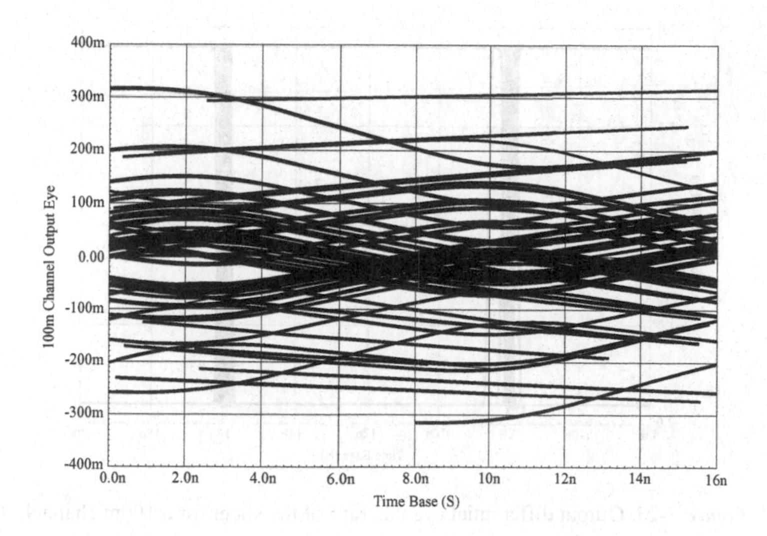

Figure 7-21. Output differential eye diagram of the 100m channel.

Figure 7-21 shows the differential eye diagram of the output of the 100m transmission channel. As shown in figure, the eye is completely closed making it impossible to receive the data without equalization. Figure 7-22 shows the eye diagram of the differential output of the FE. Again as shown in figure, the effect of the FE was to open up the vertical and horizontal eye openings of the received data in order to enable the slicer to resolve it.

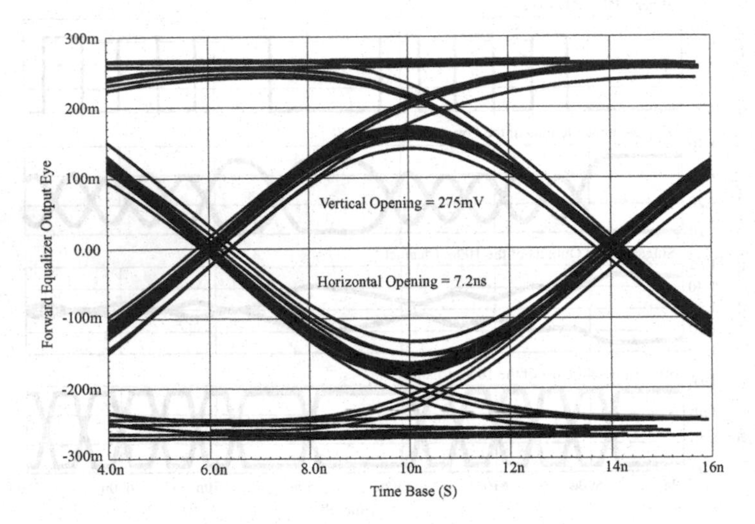

Figure 7-22. Output differential eye diagram of the FE for a 100m channel.

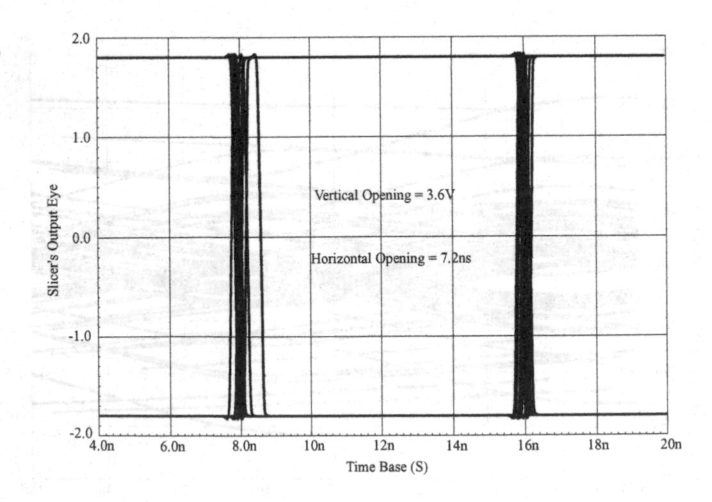

Figure 7-23. Output differential eye diagram of the slicer for a 100m channel.

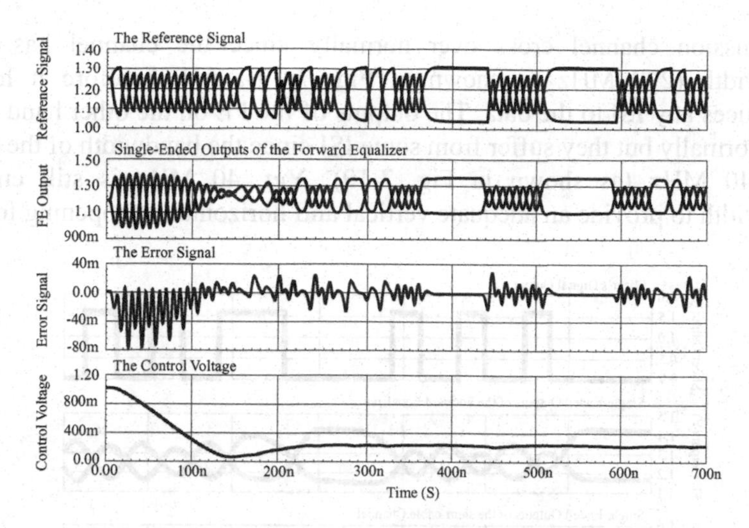

Figure 7-24. Transient response of the error and control signals for a 100m channel.

The FE achieves a horizontal opening of 7.2ns, which is only 800ps less than the ideal opening (8ns), while the vertical opening is 275mV (almost half the amplitude of the transmitted data), which is an enough overdrive for the slicer to resolve the data. Figure 7-23 shows the differential eye diagram of the output of the slicer, which is similar to the output of the FE (essentially the same data) with the exception of its digital rail-to-rail nature. Note that the vertical opening is double the supply voltage since the eye is differential.

In order to asses the adaptive behavior of the equalizer, Fig. 7-24 shows the reference signal, the FE output, the error signal, and the control voltage (used to adapt the FE). As shown in figure, at the beginning of the reception, the control voltage starts at a high value (around 1V) to guarantee enough initial boosting, this essentially causes the output of the forward equalizer to have a much faster rise and fall time than the reference signal. Therefore, the error signal will have an average negative value, which in turn causes the control voltage to drop, and consequently reduces the amount of boosting introduced by the FE. Once the output of the FE converges to the same rise and fall time as the reference signal, the error signal averages out to 0, and the control voltage starts to hold its value that assures an optimum boosting. In the 100m case, this optimum value of the control voltage is around 236mV, while its convergence time is around 300ns.

Transient simulations were also performed using the short-cable transmission channel. For this case, Fig. 7-25 shows a sample of the input data stream to the channel, the output of the channel, the output of the FE, and the final digital output of the slicer. In this case, the outputs of the transmission channel cross over normally since the channel has wide bandwidth (279 MHz as shown in Fig. 7-19), and therefore it hardly introduces any ISI to the data. The outputs of the FE on the other hand cross over normally but they suffer from some ISI since the bandwidth of the FE is only 40 MHz (as shown in Fig. 7-19). Yet, 40 MHz is still enough bandwidth to provide an adequate vertical and horizontal eye opening for the

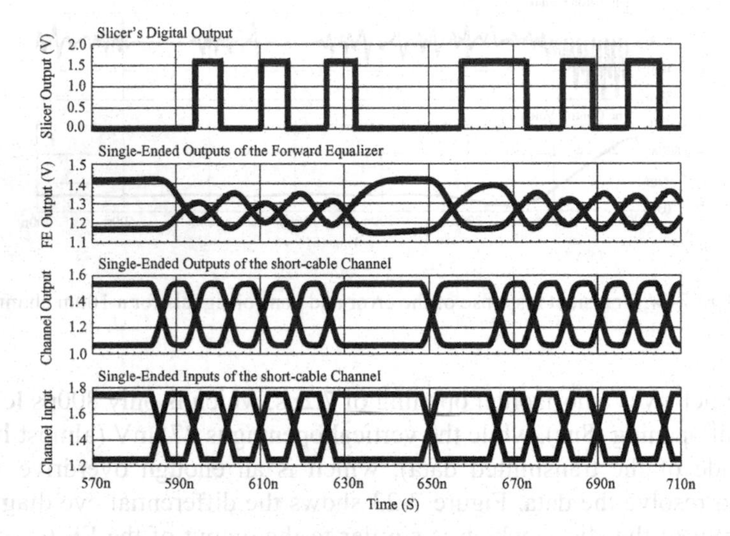

Figure 7-25. Transient response of the short-cable channel, the FE, and the slicer.

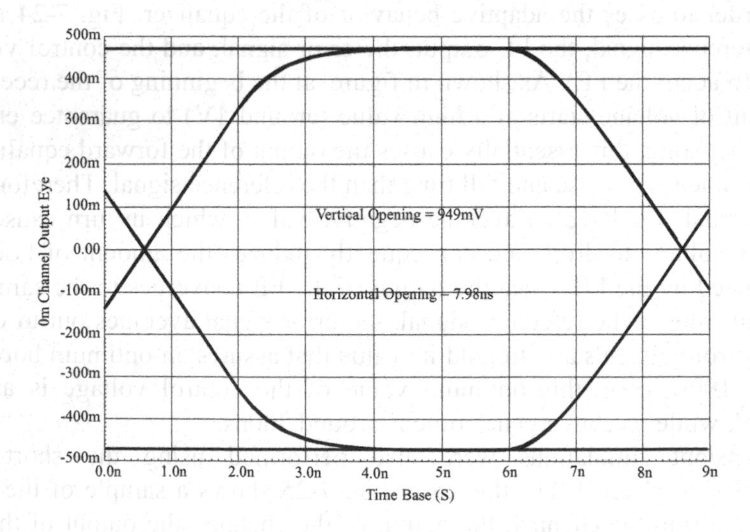

Figure 7-26. Output differential eye diagram of the short-cable channel.

slicer to receive data adequately at125 Mbps. The output of the FE is then passed to the slicer to generate the final digital output. Figure 7-26 shows the differential eye diagram of the output of the short-cable transmission channel. As expected, the eye is completely open due to the wide bandwidth of the channel, note that the vertical opening represents the peak to peak differential amplitude. Figure 7-27 shows the eye diagram of the FE output.

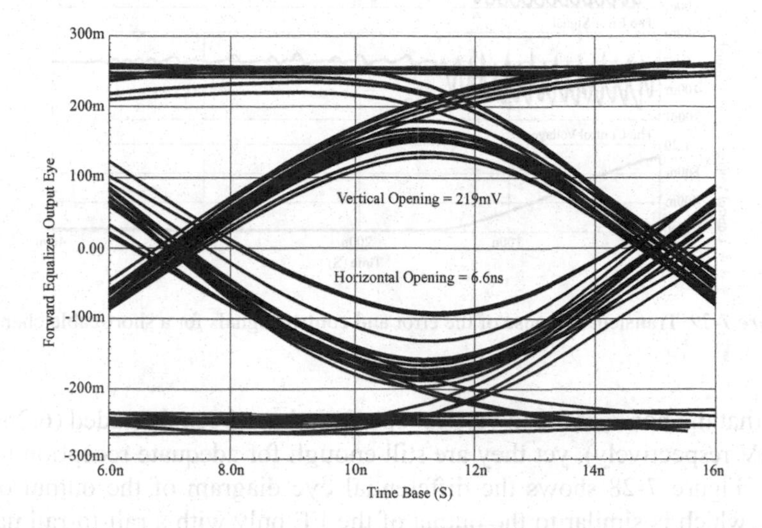

Figure 7-27. Output differential eye diagram of the FE for a short-cable channel.

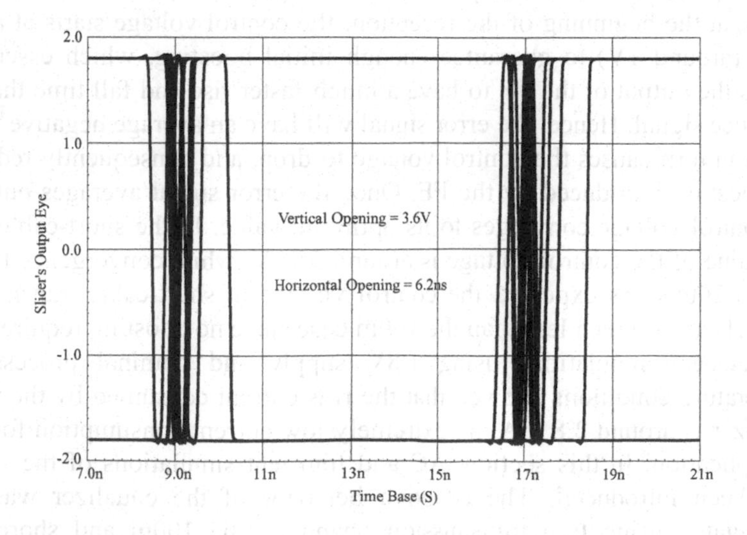

Figure 7-28. Output differential eye diagram of the slicer for a short-cable channel.

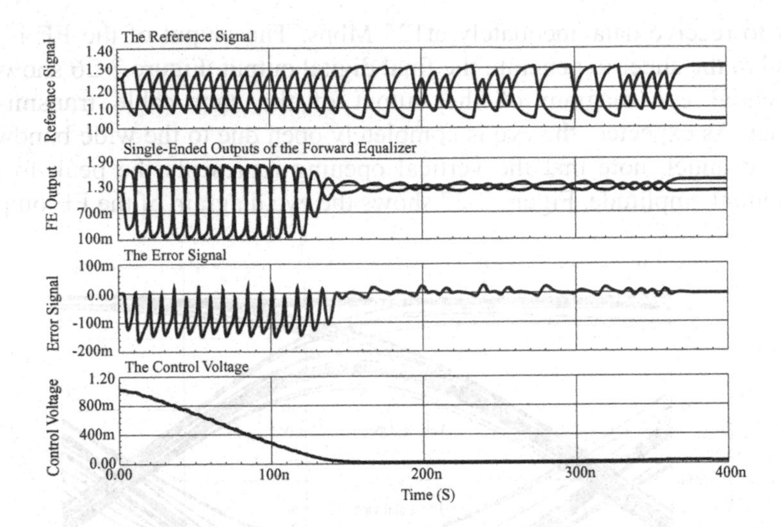

Figure 7-29. Transient response of the error and control signals for a short-cable channel.

Note that the horizontal and vertical openings have been degraded (6.2ns and 219mV respectively), yet they are still enough for adequate reception by the slicer. Figure 7-28 shows the differential eye diagram of the output of the slicer, which is similar to the output of the FE only with a rail-to-rail nature. In order to asses the adaptive behavior of the equalizer, Fig. 7-29 shows the reference signal, the FE output, the error signal, and the control voltage. Again, at the beginning of the reception, the control voltage starts at a high value (around 1V) to guarantee enough initial boosting, which essentially causes the output of the FE to have a much faster rise and fall time than the reference signal. Hence, the error signal will have an average negative value, which in turn causes the control voltage to drop, and consequently reducing the boosting introduced by the FE. Once the error signal averages out to 0, the control voltage converges to its optimum value. In the short-cable case, this value of the control voltage is around 35mV, while convergence time is around 200ns. As expected the control voltage in short-cable transmission channel case is much less than the 100m case since no boosting required.

Transient simulations using 1.8V supply and nominal process and temperature conditions showed that the rms current consumed by the whole equalizer is around 2.8 mA; an extremely low current consumption for such an application. In this section, AC and transient simulations of the system have been introduced. The adaptive behavior of the equalizer was also investigated using two transmission channels (the 100m and short-cable channels). In the next section, lab measurement results will be introduced.

The operation of the equalizer with different cable lengths will be measured and evaluated through eye diagrams at the input and output of the system.

7.5 Measurements Results

As explained in the beginning of this chapter, the equalizer was implemented as a part of a larger transceiver/repeater system that was designed to enable two high-speed IEEE 1394 transceivers to communicate

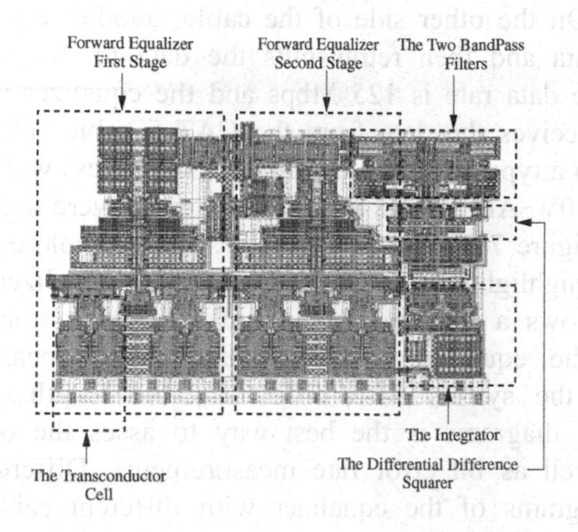

Figure 7-30. Layout of the equalizer.

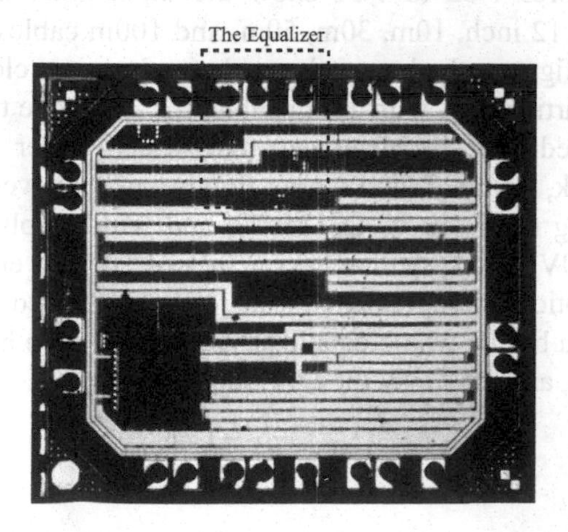

Figure 7-31. Chip micrograph showing the location of the equalizer on the chip.

Table 7-1. Equalizer's differential output eye horizontal opening.

Cable Length	Horizontal Eye Opening (ideal is 8ns)
12 inches	5.64ns
10 meters	5.39n
30 meters	7.14ns
50 meters	6.25ns
100 meters	5.64ns

across UTP CAT-5 cables that have variable length of up to 100 meters. The transceiver/repeater essentially receives the data from one high-speed transceiver (usually through metal traces) and retransmits this data across the CAT-5 cable. On the other side of the cable, another transceiver/repeater receives the data and then retransmits the data to the other high-speed transceiver. The data rate is 125 Mbps and the equalizer was used as the receiver that receives the data from the CAT-5 cable. The equalizer was implemented on a typical 180nm digital CMOS process with a supply range from 1.6V to 2.0V. All devices used in the design were digital low-voltage core devices. Figure 7-30 shows the layout of the whole equalizer with its different parts highlighted. The area of the whole equalizer is 27738 μm^2. Figure 7-31 shows a chip micrograph of the whole transceiver/repeater system with the equalizer part highlighted. Lab measurements were performed on the system with different cable lengths. As mentioned previously, eye diagrams is the best way to asses the operation of the equalizer, as well as bit error rate measurements. Differential input and output eye diagrams of the equalizer with different cable lengths were measured and table 7-1 summarizes the horizontal opening of the equalizer's output eye. Figures 7-32 to 7-36 show the input and output eyes of the equalizer with a 12 inch, 10m, 30m, 50m, and 100m cables respectively. As shown in those figures, the longer the cable is, the more closed the received eye becomes. Particularly, with 100m cable, the input eye to the equalizer is completely closed. The equalizer was also tested under different process conditions (weak, nominal, and strong conditions), as well as temperature extremes ranging from -40 °C to 125 °C, and with supply voltage varying from 1.6V to 2.0V. The system always achieved 10^{-12} bit error rate or better. Power consumption measurements were also performed and the average power consumed by the whole equalizer was found to be between 3.68 mW (at 1.6V supply), and 10.3 mW (at 2.0V supply).

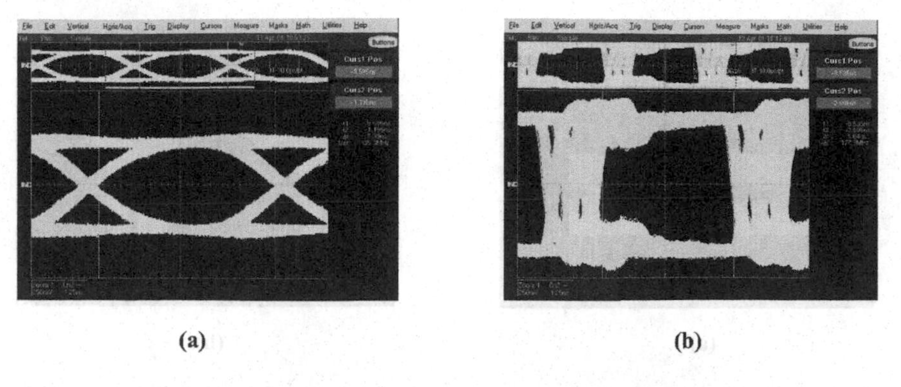

(a) **(b)**

Figure 7-32. Equalizer's eye diagrams with a 12-inche cable: (a) Input eye, (b) Output eye

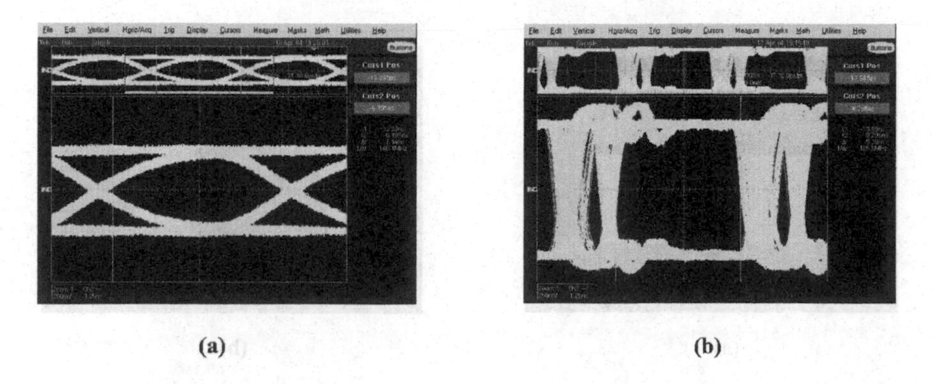

(a) **(b)**

Figure 7-33. Equalizer's eye diagrams with a 10-meter cable: (a) Input eye, (b) Output eye

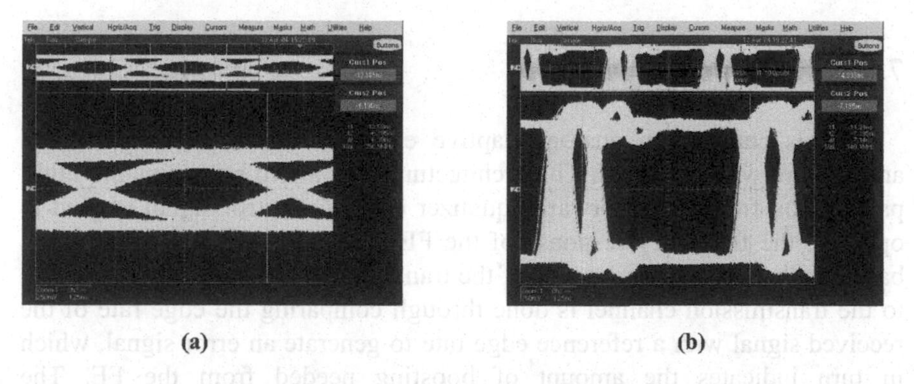

(a) **(b)**

Figure 7-34. Equalizer's eye diagrams with a 30-meter cable: (a) Input eye, (b) Output eye

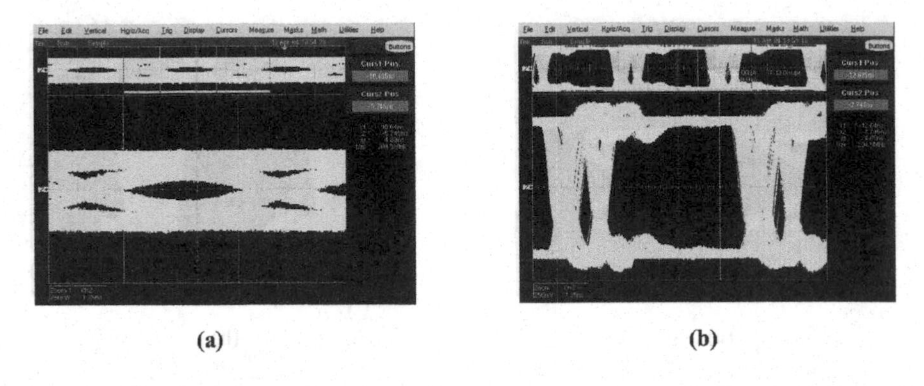

(a) (b)

Figure 7-35. Equalizer's eye diagrams with a 50-meter cable: (a) Input eye, (b) Output eye

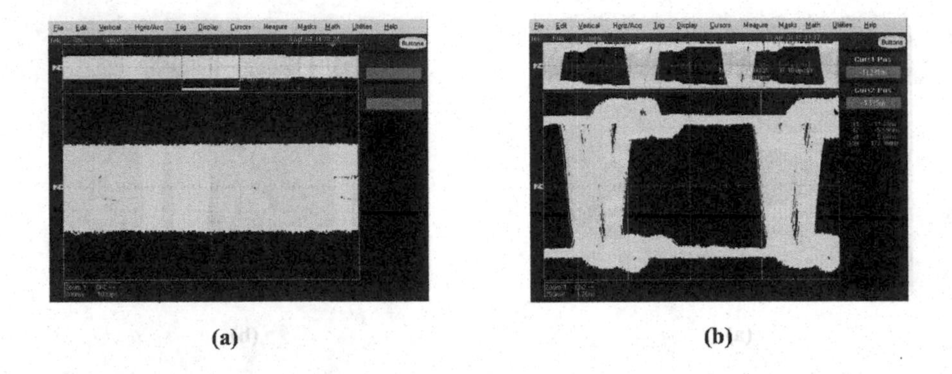

(a) (b)

Figure 7-36. Equalizer's eye diagrams with a 100-meter cable: (a) Input eye, (b) Output eye

7.6 Summary

In this chapter, an analog adaptive equalizer based on feed-forward architecture was presented. The architecture uses a two stage tunable high-pass analog filter as its forward equalizer (FE). A control signal is used to optimize the frequency response of the FE by adjusting its poles locations based on the frequency response of the transmission channel. The adaptation to the transmission channel is done through comparing the edge rate of the received signal with a reference edge rate to generate an error signal, which in turn indicates the amount of boosting needed from the FE. The architecture does not require a training signal to optimize the FE since the optimization is done based on the edge rate of the received data rather than a specific bit pattern. The implementation of the equalizer for long haul wire line communication systems was presented. Specifically, the adaptive equalizer was used for a wire line digital transceiver that operates at 125 Mbps over a variable length of up to 100 meters of an UTP CAT-5 Ethernet cable.

LIST OF REFERENCES

1. Srikanth Gondi, Randall Geiger, Jin Liu, Juergen Bareither, Scott Sterrantino, Erik Pace, "A 2V Low-Power CMOS 125baud Repeater Architecture for UTP5 Cables," *Proceedings of the Europian Solid-State Circuits Conference*, Sept. 2002.
2. IEEE Computer Society, "IEEE Standard for a High Performance Serial Bus," IEEE Inc., New York, 1995.
3. Compaq Computer Corporation, Hewlett-Packard Company, Intel Corporation, Lucent Technologies Inc, Microsoft Corporation, NEC Corporation, Koninklijke Philips Electronics N.V, "Universal Serial Bus Specifications," Revision 2.0, Draft 0.9, Dec. 21, 1999.
4. Don Anderson, "Universal Serial Bus System Architecture," Addison-Wesley, Massachusetts, 1999.
5. David M. Pozar, "Microwave Engineering," Addison-Wesley publishing company, New York, 1990.
6. Matthew N. O. Sadiku, and Lawrence C. Agba, "A Simple Introduction to the Transmission-Line Modeling," *IEEE Trans. Circuits Syst. II*, vol. 37, No. 8, pp. 991-999, Aug. 1990.
7. David A. Johns, Ken Martin, "Analog Integrated Circuit Design," John Wiley & Sons, New York, 1997.
8. Mohammed Ismail, Terri Fiez, "Analog VLSI Signal and Information Processing," McGraw-Hill, New York, 1994.
9. M. Bazes, "Two novel fully complementary self-biased CMOS differential amplifiers," *IEEE Journal of Soild-State Circuits*, vol. 26, pp. 165-168, Feb. 1991.

Bibliography

1. H. Conrad, "2.4 Gbit/s CML I/Os with integrated line termination resistors realized in 0.5/spl mu/m BiCMOS technology," *Proceedings of the Bipolar/BiCMOS Circuits and Technology Meeting,* pp. 120-122, Sept. 1997.
2. T.J. Gabara, "On-chip terminating resistors for high-speed ECL-CMOS interfaces," *Proceedings of the Fifth Annual IEEE International ASIC Conference and Exhibit,* pp. 292-295, Sept. 1992.
3. D.R. White, K. Jones, J.M. Williams, I.E. Ramsey "A simple resistance network for calibrating resistance bridges," *IEEE Transactions on Instrumentation and Measurement,* vol. 46, pp. 1068-1074, Oct. 1997.
4. I. Novak, "Modeling, simulation, and measurement considerations of high-speed digital buses," *Instrumentation and measurement Technology Conference,* pp. 1068-1074, May. 1992.
5. David A. Johns, Ken Martin, "Analog Integrated Circuit Design," John Wiley & Sons, New York, 1997.
6. R. Jacob, Harry W. Li, David E. Boyce, "CMOS Circuit Design, Layout, and Simulation," IEEE Press Series on Microelectronic Systems, New York, 1998.
7. Yuan Taur, Tak H. Ning, "Fundamentals Of Modern VLSI Devices," Cambridge University Press, Cambridge 1998.
8. R. A. Blauschild, "An Integrated Time Reference," *ISSCC Dig. Tech. Papers*, PP. 56-57, Feb. 1994.
9. Adel Sedra, Kenneth Smith "Microelectronic Circuits," Third Edition, Saunders College Publishing, Toronto 1991.
10. K. Bult and G. J. G. M. Geelen, "An Inherently Linear and Compact MOST-Only Current-Division Technique," *IEEE Journal of Solid-State Circuits*, vol. 27, No. 6, PP. 1730-1735, Dec. 1992.
11. Mohammed Ismail, Terri Fiez, "Analog VLSI Signal and Information Processing," McGraw-Hill, New York, 1994.
12. H. Wallinga and K. Bult, "Design and Analysis of CMOS Analog Processing Circuits by Means of a Graphical MOST Model," *IEEE Journal of Solid-State Circuits*, vol. 24, No. 3, PP. 672-680, Jun. 1989.

13. N. I. Khachab and M. Ismail, "Linearization techniques for n^{th}-order sensor models in MOS VLSI technology," *IEEE Trans. Circuits. Syst.*, vol. 38,PP. 1439-1450, Dec. 1991.
14. M. Ismail and D. Rubin, "Improved circuits for the realization of MOSFET-capacitor filters," *IEEE Int. Symp. Circuits. Syst.*,PP. 1186-1189, May. 1986.
15. S. T. Dupuie and M. Ismail, "High frequency CMOS transconductors," *in Analog IC Design: the current-mode approach* (C. Toumazou, F.J. Lidgey, and D. G. Haigh, eds), ch. 5, London: Peter Peregrinus Ltd., 1990.
16. P. E. Allen and D. R. Holberg, "CMOS Analog Circuit Design," Holt, Rinehart and Winston, 1987.
17. M. Ismail, "Four-transistor continuous-time MOS transconductor," *Electronics Letters*, vol. 23, PP. 1099-1100, Sept. 1987.
18. P. Ryan and D. G. Haigh, "Novel fully-differential MOS transconductor for integrated continuous-time filters," *Electronics Letters*, vol. 23, PP. 742-743, Jul. 1987.
19. D. R. Welland, S. M. Phillip, Ka. Y. Leung, G. T. Tuttle, S. T. Dupuie, D. R. Holberg, R. V. Jack, N. S. Sooch, K. D. Anderson, A. J. Armstrong, R. T. Behrens, W. G. Bliss, T. O. Dudley, W. R. Foland, N. Glover, L. D. King, "A digital read/write channel with EEPR4 detection," *ISSCC Dig. Tech. Papers*, PP. 276-277, Feb. 1994.
20. M. C. H. Cheng and C. Toumazou, "Linear composite MOSFETS (COMFET)," *Electronics Letters*, PP. 1802-1804, Sept. 1991.
21. E. Seevinck and R. F. Wassenaar, "A versatile CMOS linear transconductor/square-law function circuit," *IEEE Journal of Solid-State Circuits*, vol. SC-22, PP. 366-377, Jun. 1987.
22. S. C. Huang and M. Ismail, "Linear tunable COMFET transconductors," *Electronics Letters*, vol. 29, PP. 459-461, Mar. 1993.
23. Ayman A. Fayed, "Highly-Linear, Wide-Input-Range, Wide Control-Range, Low-Voltage Differential Voltage Controlled Transconductor," US Patent No. 6724258, April 2004.
24. Ayman Fayed, M. Ismail, "A Low-Voltage, Highly-Linear, Voltage-Controlled Transconductor," *IEEE Trans. Circuits Syst. II*, Vol. 52, No. 12, Dec. 2005.
25. Compaq Computer Corporation, Hewlett-Packard Company, Intel Corporation, Lucent Technologies Inc, Microsoft Corporation, NEC Corporation, Koninklijke, Philips Electronics N.V, "Universal Serial Bus Specifications," Revision 2.0, Draft 0.9, Dec. 21, 1999.
26. Don Anderson, "Universal Serial Bus System Architecture," Addison-Wesley, Massachusetts, 1999.
27. D. Yaklin, "Offset Comparator with Common Mode Voltage Stability," US Patent 5517134, May 1997.
28. A.L. Coban, P.E. Allen, "A 1.75 V rail-to-rail CMOS op amp," *IEEE International Symposium on Circuits and systems*, vol. 5, pp. 497-500, May. 1994.
29. Ayman A. Fayed, and M. Ismail "A High Speed, Low Voltage CMOS Offset Comparator*," Int. J. of Analog Integrated Circuits and Signal Processing,* vol. 36, No. 3, pp. 267-272, Sept. 2003.
30. Ayman A. Fayed, "High speed offset comparator," *US Patent 6400219*, June. 2002.
31. C. Galup-Montoro, M.C. Schneider, I.J.B. Loss, "Low output conductance composite MOSFET's for high frequency analog design," *IEEE International Symposium on Circuits and systems*, vol. 5, pp. 783-786, May. 1994.
32. T. Voo, C. Toumazou, "Tunable current mirror technique for high frequency analog design," *IEE Colloquium on Analogue Signal Processing,* pp. 3/1-314, 1994.

33. Richard C. Jaeger, "Introduction To Microelectronics Fabrication," Volume V, Modular Series On Solid State Devices, Addison-Wesley Publishing Company, 1988.
34. Frode Larsen, Mohammed Ismail, and Christopher Abel, "A Versatile Structure for On-Chip Extraction of Resistance Matching properties," *IEEE Transactions on Semiconductor Manufacturing*, vol.9, No. 2, PP. 281-285, Feb. 1988.
35. J. N. Burgharts, M. Soyeur, K. A. Jenkins, M. Kies, M. Dolan, K. J. Stein, J. Malinowski, and D. L. Harame, "Integrated RF components in a SiGe bipolar technology," *IEEE Journal of Solid-State Circuits*, vol. 32, PP. 1440-1445, Sept. 1997.
36. Hirad Samavati, Ali Hajimiri, Arvin R. Shahani, Gitty N. Nasserbakht, and Thomas H. Lee, "Fractal Capacitors," *IEEE Journal of Solid-State Circuits*, vol. 33, No. 12, PP. 2035-2041, Dec. 1998.
37. O.E. Akcasu, "High capacitance structures in a semiconductor device," U.S. Patent No. 5208725, May 1993.
38. E. Pettenpaul, H. Kapusta, A. Weisgerber, H. Mampe, J. Luginsland, and I. Wolff, "Models of lumped elements on GaAs up to 18 GHz," *IEEE Transactions on Microwave Theory Tech.*, vol. 36, PP. 294-304, Feb. 1988.
39. T. J. Gabara, S. C. Knauer, "Digitally adjustable resistors in CMOS for high-performance applications," *IEEE J. Solid-State Circuits*, vol. 27, pp. 1176-1185, Aug. 1992.
40. A. DeHon, T. Knight, Jr., T. Simon, "Automatic impedance control," *IEEE International Solid-State Circuits Conference*. Digest of Technical Papers, pp. 164-165, 283, Feb. 1993.
41. Thaddeus Gabara, Wilhelm Fischer, Wayne Werner, Stefan Siegel, Makeshwar Kothandaraman, Peter Metz, and Dave Gradl, "LVDS I/O Buffers with a Controlled Reference Circuit," *Proceedings of the 10th Annual IEEE International ASIC conference*, pp. 311-315, Sept. 1997.
42. Hongjiang Song, "Dual mode transmitter with adaptively controlled slew rate and impedance supporting wide range data rates," *Proceedings of the 14th Annual IEEE International ASIC/SOC conference*, pp. 321-324, Sept. 2001.
43. M. Ismail, S.V. Smith, and R.G. Beale, "A new MOSFET-C universal filter structure for VLSI," *IEEE J. Solid-State Circuits*, vol. SC-23, pp. 183-194, Feb. 1988.
44. S. Sakurai, and M. Ismail, "A CMOS square-law programmable floating resistor independent of the threshold voltage," *IEEE Trans. Circuits Syst. II*, vol. 39, pp. 565-574, Aug. 1992.
45. R. Schaumann, M. S. Ghausi, and K. R. Laker, "Design of Analog filters, passive, active RC, and switched-capacitor", Prentice-Hall, Englewood Cliffs, NJ, 1990.
46. Satoshi Sakurai, Mohammed Ismail, Jean-Y ves Michael, Edgar Sanchez-Sinencio, Robert Brannen "A MOSFET-C Variable Equalizer Circuit with Simple On-Chip Automatic Tuning," *IEEE Journal of Solid-State Circuits*, vol. 27, NO. 6, June 1992.
47. K. Nagaraj, "New CMOS floating voltage-controlled resistor," *Electronics Letters*, vol. 22, PP. 667-668, 1986.
48. S. P. Singh, J. V. Hanson, and J. Vlach, "A new floating resistor for CMOS technology," *IEEE Trans. Circuits. Syst.*, vol. 36, PP. 1217-1220, Sept. 1989.
49. M. Steyaert, J. Silva-Martinez, and W. Sansen, "High-frequency saturated CMOS floating resistor for fully-differential analog signal processors," *Electronics Letters*, vol. 27, PP. 1609-1611, 1991.
50. Behzad Razavi, "Design of Analog CMOS Integrated Circuits," McGraw-Hill, New York, 2001.
51. Ayman A. Fayed, and M. Ismail "A Digital Tuning Algorithm For On-Chip Resistors," *Proceedings of the 2004 International Symposium on Circuits and Systems,* Vol. 1, pp. 936-939, May 2004.

52. Ayman A. Fayed, and M. Ismail "A Digital Calibration Algorithm for implementing accurate On-Chip Resistors," *Int. J. of Analog Integrated Circuits and Signal Processing*, Accepted, Nov. 2005.
53. Kyoung-Hoi Koo, Jin-Ho Seo, Myeong-Lyong Ko, and Jae-Whui Kim, "Digitally tunable on-chip resistor in CMOS for high-speed data transmission," *ISCAS 2003. Proceedings of the 2003 International Symposium on Circuits and Systems*, vol. 1, pp. 185-188, May 25-28, 2003.
54. U. S. H. Qureshi, "Adaptive Equalization," *Proc. IEEE* 1985.
55. Simon Haykin, "Adaptive Filter Theory," Fourth Edition, Prentice-Hall, New Jersey, 2002.
56. J. G. Proakis, "Digital Communications," Fourth Edition, McGraw-Hill, New York, 2001.
57. Herbert Taub, Donald L. Schilling, "Principles Of Communication Systems," Second Edition, McGraw-Hill, New York, 1992.
58. Bang-Sup Song, David C. Soo, "NRZ Timing Recovery Technique for Band-Limited Channels," *IEEE Journal of Solid-State Circuits*, vol. 32, NO. 4, April 1997.
59. Michael Q. Le, P. J. Hurst, J. P. Keane "An Adaptive Analog Noise-Predictive Decision-Feedback Equalizer," *IEEE Journal of Solid-State Circuits*, vol. 37, NO. 2, Feb. 2002.
60. Jae-Yoon Sim, Jang-Jin Nam, Young-Soo Sohn, Hong-June Park, Chang-Hyun Kim, Soo-In Cho "A CMOS Transceiver for DRAM BUS System with a Demultiplexed Equalization Scheme," *IEEE Journal of Solid-State Circuits*, vol. 37, NO. 2, Feb. 2002.
61. D. Lin, "High bit rate digital subscriber line transmission with noise-predictive decision-feedback equalization and block coded modulation," *Proc. IEEE Int. Conf. Communications*, pp. 17.2.1-17.3.5, 1989.
62. R. Wiedmann, J. Kenney, and W. Kolodziej "Adaptation of an all-pass equalizer for DFE," *IEEE Trans. Magn.*, vol. 35, pp. 1083-1090, Mar. 1999.
63. N. Garrido, J. Franca, and J. Kenney "A Comparative study of two adaptive continuous-time filters for decision feedback equalization read channels," *Proc. IEEE Int.Symp. Circuits and Systems*, pp. 89-92, 1997.
64. P. McEwen, and J. Kenney "All-pass forward equalizer for decision feed-back equalization," *IEEE Trans. Magn.*, vol. 31, pp. 3045-3047, Nov. 1995.
65. E. Lee, D. Messerschmitt, "Digital Communication," Kluwer, pp. 159-167, Norwell, MA 1988.
66. Simon Haykin, "Digital Communication," John Wiley & Sons, New York, 1988.
67. Srikanth Gondi, Randall Geiger, Jin Liu, Juergen Bareither, Scott Sterrantino, Erik Pace, "A 2V Low-Power CMOS 125baud Repeater Architecture for UTP5 Cables," *Proceedings of the Europian Solid-State Circuits Conference*, Sept. 2002.
68. IEEE Computer Society, "IEEE Standard for a High Performance Serial Bus," IEEE Inc., New York, 1995.
69. David M. Pozar, "Microwave Engineering," Addison-Wesley publishing company, New York, 1990.
70. Matthew N. O. Sadiku, and Lawrence C. Agba, "A Simple Introduction to the Transmission-Line Modeling," *IEEE Trans. Circuits Syst. II*, vol. 37, No. 8, pp. 991-999, Aug. 1990.
71. M. Bazes, "Two novel fully complementary self-biased CMOS differential amplifiers," *IEEE Journal of Soild-State Circuits*, vol. 26, pp. 165-168, Feb. 1991.
72. Ayman A. Fayed, "Adaptive Techniques for Analog and Mixed Signal Integrated Circuits," Ph.D. Thesis, The Ohio State University, Fall 2004.

Index